SHARPENING
THE COMPLETE GUIDE

SHARPENING
THE COMPLETE GUIDE

JIM KINGSHOTT

Guild of Master Craftsman Publications

First published 1994 by
Guild of Master Craftsman Publications Ltd,
166 High Street, Lewes, East Sussex BN7 1XU
© Jim Kingshott 1994

ISBN 0 946819 48 3

Illustrations © Jim Kingshott 1994

Photography © Jim Kingshott 1994

Designed by Fineline Studios
Printed and bound in Great Britain
by the Bath Press

To my son
John

Contents

INTRODUCTION

When asked to write a book about sharpening, I approached the task with some trepidation. Over the years I have heard many disputes in the workshop about this controversial subject. I could see in my mind's eye the piles of letters from readers who disagreed with what I had written.

Most craft knowledge is passed on to us from previous generations, either orally or in written form. There are, from time to time, innovations, and a few are permanently adopted by the trade. Most of these innovations are gimmicky; once thoroughly tried and found wanting, they are rejected. When one considers that the ancient Egyptians worked wood by hand in 3000BC, using methods similar to those we use today, it is easy to appreciate that most things will have been tried. This constant sifting and refining of techniques - including sharpening techniques - by those craftsmen who have gone before us, leaves us indebted to them. For without sharp tools it is impossible to produce anything worthwhile from wood. Over the years many different methods and materials have been used to achieve the best edge possible. We are in the fortunate position to benefit from this bed of knowledge gained by the craft over a period of 5000 years.

It was after considering these things that I decided to pass on some of what I have picked up regarding sharpening during my time in the trade. I do not say these are the only methods, but they are those which I consider worth knowing.

Sharpening is the most important skill that a woodworker must acquire, yet scant attention is paid to it in our colleges. Sharpening is looked upon by many in the craft as a necessary evil, which wastes production time. Yet sharp tools are a pleasure to use; blunt tools are an abomination. Work produced by sharp tools has a crisp, clean appearance while blunt tools leave a dull, coarse surface. The precision of cut from a sharp tool enables one to produce perfect joints. The skill of sharpening, as with other skills, is developed by practice. Do not be disappointed when trying new methods if they do not work out too well at first: 'Practice makes perfect'.

I hope that you will gain something from this book; it should help to make your tools sharper, and consequently allow you to enjoy working wood that much more. Remember, life is like a grindstone: some people it grinds down, others it just polishes.

Please forgive me for using the masculine pronoun throughout this book, and for using 'craftsman' instead of that horrid word 'craftsperson'. I mean no slight to women; in all cases I refer to either sex.

CHAPTER 1

WHAT IS SHARPNESS?

A DEFINITION

What do we mean when we say that something is 'sharp'? At what point do we say that an edge that was sharp is now 'blunt'? If we are to talk about sharpening it is important that we can define these two important words. The *Oxford English Dictionary* gives the following definitions: **Sharp**, having a thin cutting edge or a fine point; well adapted for cutting or piercing. **Blunt**, having an obtuse, thick, or dull edge or tip; rounded; not sharp.

Neither of these definitions helps us very much. The truth is, there is no objective measure of sharpness. One man's sharp is another man's blunt. The quickest way to start an argument in a workshop is to get the craftsmen talking about sharpening. It is probably the most controversial of all woodworking subjects.

Throughout this book the word 'sharp' is used to describe the cutting edge of a tool that is suitable for the work it is intended to perform; and the word 'blunt' will be used to describe an edge that needs sharpening.

I state above that there is no way of measuring sharpness. It may interest you to know, before the Samurai were outlawed in Japan in 1868, the swordsmiths had found a way of classifying the sharpness of their blades. A number was engraved on the handle of a sword; this was the number of human bodies the sword had cut through in one stroke. Even if we could find a way of applying this to woodworking tools I don't think it would be very acceptable!

Fig 1.1 Looking for the candle. A jeweller's lens is very useful.

THE CANDLE

As there is no way of measuring sharpness it is very difficult for the beginner to know when a tool is sharp. Some people feel the edge with the ball of the thumb. Do not do this; if the edge is really sharp it will cut you. The only way to inspect an edge for sharpness is as follows: hold the tool with the edge upwards, in a good light. Tip the tool slightly back and forth. What you are looking for is a fine white line along the cutting edge. This is called the candle, and it indicates bluntness. It shows up quite plainly when it catches the light. Some craftsmen use a magnifying glass when looking for the candle (*see* Fig 1.1). The first time you look for this feature, use a tool that you know is blunt. Once you know what you are looking for it is easy to see it. The thicker the candle line the blunter the tool. So we can now define 'sharp' as an edge that has no candle. Conversely, a blunt tool is obviously one that displays this feature.

Let us look at an edge in some detail. It consists of two flat surfaces that meet at the cutting edge. The angle between these two surfaces has an effect on the cutting ability of the tool. In most cutting operations the tool is entering the wood. An edge with a large angle between the two surfaces requires more effort to push it into the wood than one with a small angle (*see* Fig 1.2). However the edge with a small angle is weak and does not last. Because of this, the angles used to obtain a cutting edge are a compromise.

HORSES FOR COURSES

A tool can be sharpened to suit the job in hand. For instance, when using a soft, mild timber the angle can be small. This will give an edge that requires less effort to use than one with a steep

Fig 1.2 Angle of tool entering wood. The low-angled blade at A cuts with less effort than the steep-angled blade at B.

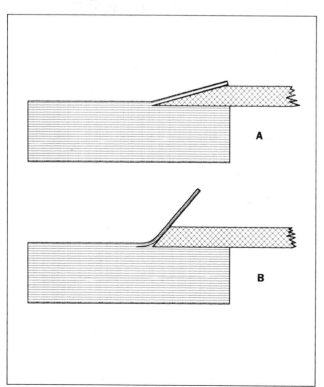

Fig 1.3 Angle of blade pitch in plane, as opposed to sharpening angle. A is the angle seen by the wood being planed, B is the pitch of the iron, and C is the sharpening angle.

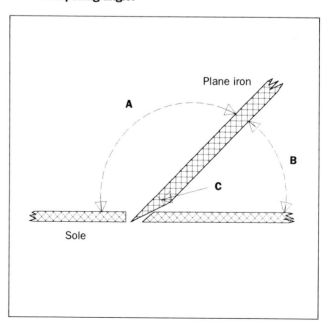

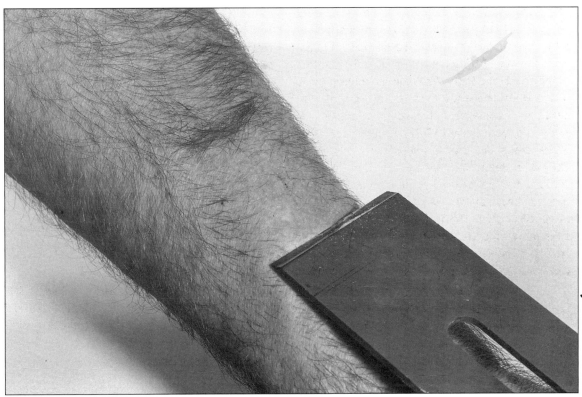

Fig 1.4
Shaving the
hairs from
my arm with
a sharp
plane iron.

angle. However, it would be unsuitable for use on a dense, hard wood. The tool would blunt very quickly, and the times between sharpening would be impractically short.

This brings me to a very controversial point. There are two opposing opinions regarding the sharpening angle of plane blades. The first states that the smaller the angle of the cutting edge the easier the tool cuts. This concurs with what has already been said. However, there is another theory, just as valid. This states that the only angle that the wood sees is the pitch of the blade (*see* Fig 1.3). The effect that the sharpening angle has is to provide clearance behind the cutting edge. Now I think that there is a lot to be said for both arguments, and arguments there have been. I have seen craftsmen who are normally calm and rational come near to blows over the issue. My advice is to experiment, and whatever works best for you, adopt that. After all is said and done, you are the person who has to use the tool. (*See* page 56 for more on this subject.)

THE IMPORTANCE OF SHARPNESS

The first requirement, if one is to produce quality woodwork, is sharp tools. When instructing apprentices I would sharpen a plane iron and then shave the hairs from my arm with it, forbidding them to do the same on their own arms (which of course they did, the minute my back was turned). This made the point, impressed the apprentices, and left a bare patch on my arm which looked odd until the hair grew again (*see* Fig 1.4). Often I would take a tool away from a lad, sharpen it and

Fig 1.5 Toothing plane iron.
The grooves on the face cause the
cutting edge to be serrated. A scratch on the
face of a plane iron has a similar but smaller effect.

then return it to him. The difference between using a blunt tool and a sharp one was immediately appreciated. I have often been approached by DIY enthusiasts and asked to sharpen and set a plane. I have to explain to them that even under the very best circumstances the iron will only stay sharp for about half an hour's use. You will by now appreciate how important it is for a woodworker to be proficient in the art of sharpening.

If you look at the photograph of a toothing plane iron (see Fig 1.5), you will see that the edge is serrated. This is caused by the grooves on the face of the blade. (Incidentally a plane blade is

Fig 1.6 The face of the plane iron must be so smooth that it appears as if it were chromium plated. The reflection of the rule in the face of this iron shows the surface polish.

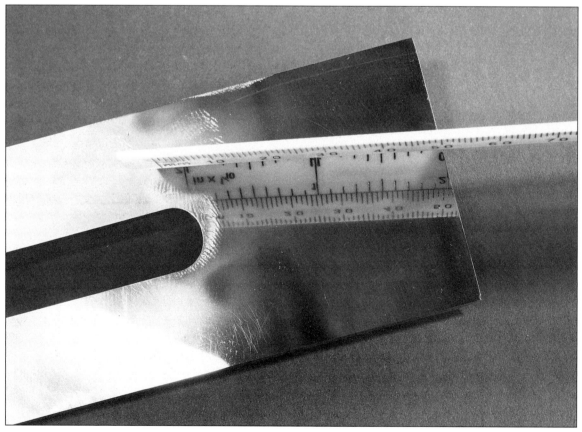

often referred to as an iron. This term dates from the days when craftsmen made their own planes and obtained the blade from the local blacksmith. Not knowing about steel they referred to all ferrous metal as iron.) If the grooves on the face of the toothing iron caused the serrated edge, it can be seen that any scratch on the face of a tool is going to produce a gap in the cutting edge. No matter how small the scratch is it will have some effect on the edge. The process of sharpening is performed by abrading the surface of the tool away, usually with a stone. This of course leaves scratches in the surface of the tool being sharpened. Therefore the final process of the sharpening operation is to polish away these scratches. The two surfaces that form the cutting edge should appear as if they have been chromium plated (see Fig 1.6).

METALLURGY

While most metals can be sharpened, few have the characteristics that we require for a cutting tool. The ancient Egyptians, who were very proficient woodworkers, did not have the benefit of steel. Some 3,000 years BC they had tools made from copper, which had been work hardened. Today we have the benefits of high technology. The characteristics of the steel from which our tools are made can be precisely controlled, and as craftsmen we should know just what characteristics we require. This will help in choosing our tools.

There are some processes and terms that need to be understood if we are to be capable of discerning the difference between steels. The following terms have particular meanings when applied to metal:

Hardness is the ability of a metal to withstand scratching, abrasion and wear, and its ability to resist penetration. This last attribute is used to measure hardness. There are several systems for specifying the hardness of metal. The main ones are Brinell, Vickers, Rockwell 'B' and Rockwell 'C'.

The last named is the one in general use for woodworking tools.

Rockwell 'C' is the system that is used throughout this book, but I have included a conversion table (see Fig 1.7). The machine used to determine the hardness in Rockwell 'C' pushes a 120° diamond cone into the surface of the metal with a known force. The machine measures the depth of the penetration – the deeper the penetration the softer the metal. A Rockwell number is then assigned to the metal – the higher the number, the harder the metal. Most tools fall between 58° and 61° Rockwell 'C'.

Brittleness is the propensity of a metal to break with little deformation.

Ductility is the ability of a material to be permanently deformed in its cold state. Ductility is dependent on strength and plasticity; it is the opposite of brittleness.

Toughness is the ability of a metal to withstand sudden heavy loading without fracture.

Today all our cutting tools, are made from steel. To make steel one must first make pig iron. Iron ore, coke, and limestone are fed into the top of a blast furnace. Hot air is forced into the bottom of the furnace, causing the coke to burn and raising the temperature to 1400°C. Pig iron, which contains three to four per cent carbon, is tapped from the bottom of the furnace.

The fundamental difference between pig iron and steel is the carbon content. The carbon content of steel ranges from 0.1 to 1.7 per cent. If more than 1.7 per cent is present, graphite carbon is formed and the material is classified as cast iron. To convert pig iron into steel it is necessary to remove some of the carbon and other impurities by oxidation. Many grades of steel are produced, but they can be divided into two main groups: **plain-carbon steels** and **alloy steels**.

Plain-carbon steels can be divided into four

Fig 1.7 Conversion table of hardness scales.

Brinell			Vickers Hardness No.	Rockwell 'B' 1/16in (1.5mm) ball	Rockwell 'C' 120° diamond cone	Scleroscope	Brinell			Vickers Hardness No.	Rockwell 'B' 1/16in (1.5mm) ball	Rockwell 'C' 120° diamond cone	Scleroscope
Diameter mm	Hardness No.	Calculated Tonnage					Diameter mm	Hardness No.	Calculated Tonnage				
2.20	782	171	1170		70	106	4.00	228	50	234	98	21	33
2.25	744	162	1050		68	100	4.05	223	49	229	97	20	32
2.30	713	155	935		66	95	4.10	217	47	223	96	18	31
2.35	683	149	865		64	91	4.15	212	46	218	96	17	31
2.40	652	142	802		62	87	4.20	207	45	213	95	16	30
2.45	627	136	756		60	84	4.25	202	44	208	94	15	30
2.50	600	131	708		58	81	4.30	196	43	202	93	13	29
2.55	578	126	670		57	78	4.35	192	42	198	92	12	28
2.60	555	121	634		55	75	4.40	187	41	193	91	10	28
2.65	532	116	598		53	72	4.45	183	40	189	90	9	27
2.70	512	112	570		52	70	4.50	179	39.5	185	89	8	27
2.75	495	108	545		50	67	4.55	174	39	180	88	7	26
2.80	477	104	521		49	65	4.60	170	38.5	176	87	6	26
2.85	460	100	500		47	63	4.65	166	38	172	86	4	25
2.90	444	97	480		46	61	4.70	163	37.5	169	85	3	25
2.95	430	94	462		45	59	4.75	159	36.5	164	84	2	24
							4.80	156	36	161	83	1	24
3.00	418	91	447		44	57	4.85	153	35	158	82		23
3.05	402	88	426		42	55	4.90	149	34	155	81		23
3.10	387	84	409		41	54	4.95	146	33.5	152	80		22
3.15	375	82	394		40	52							
3.20	364	79	382		38	51	5.00	143	33	149	79		22
3.25	351	76	366		37	49	5.05	140	32	146	78		21
3.30	340	74	352		36	48	5.10	137	31.5	143	77		21
3.35	332	72	343		35	46	5.15	134	31	140	76		21
3.40	321	70	331		34	45	5.20	131	30	137	74		20
3.45	311	68	321		33	44	5.25	128	29.5	133	73		20
3.50	302	66	311		32	43	5.30	126	29	131	72		
3.55	293	64	301		31	42	5.35	124	28.5	129	71		
3.60	286	62	294		30	40	5.40	121	28	126	70		
3.65	277	60	284	104	29	39	5.45	118	27	123	69		
3.70	269	59	276	104	28	38	5.50	116	26.5	121	68		
3.75	262	57	268	103	26	37	5.55	114	26	119	67		
3.80	255	55	261	102	25	37	5.60	112	25.5	117	66		
3.85	248	54	254	102	24	36	5.65	109	25	115	65		
3.90	241	52	247	100	23	35	5.70	107	24.5	113	64		
3.95	235	51	241	99	22	34	5.75	105	24	111	62		

grades according to their carbon content:

■ Low-carbon steel	0.1 – 0.15 per cent carbon content
■ Mild steel	0.15 – 0.3 per cent carbon content
■ Medium-carbon steel	0.3 – 0.7 per cent carbon content
■ High-carbon steel	0.7 – 1.7 per cent carbon content

Only the last grade is suitable for woodcutting tools. The hardness and tensile strength of medium and high-carbon steels can be improved by heat treatment. This, however, reduces their toughness and increases brittleness.

Alloy steels are carbon steels that contain very small amounts of other elements. The principal ones are listed below, together with the effect they have on the steel.

■ **Chromium and Molybdenum** increase hardness and strength but not brittleness. Chromium also helps resistance to corrosion. Molybdenum reduces temper brittleness and allows a metal to operate continuously at high temperatures without becoming brittle.
■ **Manganese** is added to improve the steel's mechanical properties.
■ **Vanadium** improves the elasticity of the steel.
■ **Tungsten** is a very hard element that improves the grain structure of the steel. It also confers the property of 'red hardness', which is the ability to hold an edge even at red heat.

There are other elements in steel, including very small quantities of sulphur and silicon, which it would be too expensive to remove. These also have an effect on the steel's character.

When it comes to choosing the steel that our tools are made from, we have little say in the matter. The manufacturer chooses the raw material that he considers the best for the purpose. There is also the question of cost. For instance, tungsten-tipped machine cutters are several times the price of high-speed steel cutters. Then we have the difference between an old chisel manufactured from high-carbon steel and a modern one made from alloy steel. While the modern one is tough and will hold an edge considerably longer than the old chisel, it cannot be made as sharp in the first place. I will discuss the choice of steel in detail for each tool as I describe the particular sharpening technique that applies to it.

HEAT TREATMENT

While I am not suggesting that you should undertake the heat treatment of steel yourself, it is useful to know how this affects the property of the steel. If a tool has been subjected to a high temperature – by the misuse of a high-speed bench grinder for instance – the steel will have changed. It is useful to be able to recognize this when it has happened, and be able to ascertain the extent of the damage. If it is a valued tool, knowing what is required to return it to its original state is important.

When steel is heated its temperature rises at a uniform rate until it reaches 700°C. At this point the temperature remains constant for a short time. It then continues to rise at a slower rate until it reaches 800°C, when it reverts to the original rate. When the metal is allowed to cool, this process is reversed. The point at which the temperature rise pauses when the metal is being heated is known as the **decalescence** point. The point at which the steel's temperature pauses on cooling is known as the **recalescence** point. This is the point at which the structure of the steel begins to change. The structure of the steel continues to change until it reaches an upper critical point.

This varies according to the carbon content of the steel. If the steel is allowed to cool naturally the structure of the steel reverts to its original state. However, if the rate of cooling is reduced or extended, the internal structure of the steel will remain permanently changed.

Heat treatment can be used in the following ways:

■ **Annealing** makes the material soft and ductile.

■ **Normalizing** relieves any stresses, and returns the material to its original condition.

■ **Hardening** is carried out by heating the steel to a cherry red and then cooling it suddenly (also known as quenching).

■ **Tempering** restores brittleness and some degree of toughness to hardened steel so that it is suitable for woodworking tools. This is done by reheating the material to a point well below decalescence and then quenching it.

Steel that is capable of being sharpened and holding an edge, cannot be filed. This is quite a good test of the suitability of an unknown steel. The file should skate over the surface not even marking the steel. Saw blades cannot conform to this rule as they are sharpened by filing the teeth. The steel required here is one that must be capable of being bent and staying bent, so that the teeth may be set. However, should the whole blade bend, then it must spring back straight. This is quite a tall order, and is one of the reasons why a quality saw blade is an expensive item. There are many saws sold today with inductance-hardened teeth that cannot be sharpened. These saws are aimed at the building site worker and the DIY enthusiast, who are unable or unwilling to sharpen their saws. Because these saws are made to be thrown away when they eventually become blunt, they are not made to the same precision as a top quality saw. (Saw sharpening is dealt with in Chapter Eleven.)

WHATEVER WORKS FOR YOU

You will have begun to appreciate that a sharp edge is a very important part of woodworking. It is not something that one can take for granted. Much work goes into the preparation of the tool. This requires a certain amount of expertise and knowledge, which takes time and practice to acquire. The novice should experiment with different methods, keep an open mind and be prepared to try out new ideas. One should not be too influenced by what other people state as the correct and only way to obtain a sharp edge. The method that works best for you is the one to use. Once a tool is put into tip-top condition it needs to be looked after. A very fine edge only needs the slightest knock to spoil it, and time will have to be spent returning it to that state.

CHAPTER 2

The Sharpening Process

It All Takes Time

I will admit here that I tend to treat my tools as though they were living beings. If one of them is in poor condition it worries me to such an extent that I must stop whatever I am doing and put it right. I suppose that this is understandable when one thinks of the length of time we have been together, over 50 years in some cases. Most working days they have been my constant companions, serving me well. It is only to be expected that I should feel guilty if I neglect them. A craftsman is completely useless without his tools, and the pleasure of expressing oneself in what one makes is dependent on them. In fact it would not be incorrect to say that but for the tools there would be no woodwork. Now, there's a sobering thought!

A good tool in peak condition is a pleasure to work with. A poor tool, or one out of condition, is an abomination. To get the maximum amount of enjoyment from your work, obtain the best tools you can find and keep them in the best state you are able to. Sharpening is not a minor skill of the trade. Many novices are not aware of what sharp is. The hobbyist working on his own might never be aware that his tools are not up to scratch. I hope that the descriptions below and the instructions in the following chapters will help every craftsman who is striving to obtain a sharp edge.

Some practice is needed. My first attempt at sharpening a jack plane iron stands out very clearly in my memory, even though it was many years ago. When I started my apprenticeship at the age of 14, I was given a list of tools that I should have, one of which was a jack plane. I was impatient to use my nice new tools and, for the very first week at work, I was disappointed not to be given any productive work. However, the time came when, after some instruction, I was given a length of wood to prepare. First the plane had to be sharpened. My apprentice master knocked the iron out, showing me how to hold it by rubbing it a couple of times over my oilstone. He then left me to get on with it. I must have taken that iron to him 20 times. Each time he would look at the edge and say, 'You could ride bareback to London on that and not damage yourself.' He left me sharpening all of one long morning before taking the iron and doing it himself, talking me through the whole process. It was at least a year before I managed to get an edge on my tools that was anywhere near as sharp as his.

I relate this story to show that it is not only the knowledge that is required. Skill is also needed, acquired by practice. There is also a feel for the tool as it is being honed that only comes with experience. Never be completely satisfied with the edge that you are getting on your tools. Keep striving for an improvement. A few years ago I obtained a Japanese water stone. The edge this produced on my tools was superior to that which I

Fig 2.1 The face of a plane iron that has been sharpened on a coarse stone. Note the scratches. The serrated edge is just visible.

had considered for years as the best obtainable. So you see it can happen to all of us, there is always a better way, if only we can find it.

ABRADING METAL AWAY

Every method of sharpening involves abrading metal away with some form of abrasive material. Whether we need to remove a large amount of metal or just a small amount, an abrasive is used. I will discuss these at length in later chapters, but first I wish to describe the overall sharpening process. It is important to realize that there are many different ways to reach the same end. There

are also several different types of cutting edge. Every craftsman has his own preferred type and his own way of achieving it. I will try to explain them all and you can then try each one out and find which suits you best.

First it is necessary for me to define a few more terms: **Honing** is the process of rubbing a tool on the surface of a stone to obtain a sharp edge. There are numerous stones, both natural and man-made, used for this purpose. (They are discussed fully in Chapter Four.) The **bevel** is the angle at the sharpening edge of the blade (*see* Fig 1.2). A **strop** is a strip of leather dressed with an extremely fine abrasive such as jeweller's rouge. The leather is sometimes glued to a piece of wood. Some strops are two-sided, each side being dressed with a different abrasive. **Grinding** is the process of removing metal by machine, with an abrasive wheel or belt.

In Chapter One I described the cutting edge as being produced by two sloping surfaces at an angle to one another that meet at the edge. It is this edge that does the cutting. When I was an apprentice, an old craftsman told me: 'The cutting edge is two sloping flat planes. Each in turn is rubbed on the stone, what you are after is a point where they meet that is as thin as it is possible to get it. The point is so fine that what you want there is nothing. Now then, when you have got

nothing, that's something isn't it?' Since that conversation I have often mused on the fact that most woodwork is conditional on a part of a tool which is so small that it can be described as nothing. We spend so much time and effort trying to achieve nothing.

The coarseness of the material used to sharpen a tool has an effect on the edge. Because of the size of the grains of abrasive, it is impossible to get a really sharp edge with a coarse material. The large grains of abrasive scratch the surface and break the fine edge away. A serrated edge is produced which can be seen with a powerful magnifying glass (*see* Fig 2.1).

Fig 2.2 Double bevel.

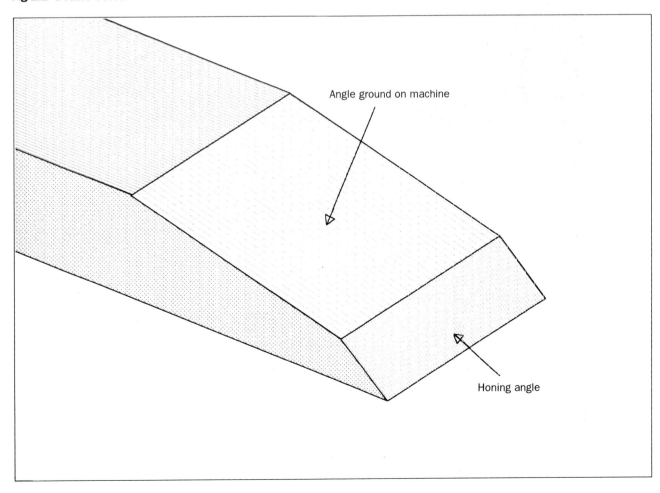

Angle ground on machine

Honing angle

However, with a new tool or one in poor condition there is a need to remove quite a lot of metal. This would be very laborious with a fine abrasive. At least two grades of abrasive will be required, a coarse one to remove the bulk of the material and a fine one to get a sharp edge.

THE DOUBLE BEVEL

When it comes to removing a fair quantity of metal from a tool, it can be a very arduous task if undertaken by hand. It is far less work, and much quicker, to remove this by grinding it away with a machine. This has led to the technique of the **double bevel** (*see* Fig 2.2). Here there is one angle produced by the grinder and another by honing. This saves much time because, when the tool needs honing, there is only a narrow band of metal to rub away. When this band becomes wide the tool is taken to the grinder again. This has been the basic technique in most workshops for many years. Originally the grinding was carried out on a large-diameter sandstone wheel turned by hand. Today most workshops have an electric

Fig 2.3 A hollow ground blade presented to the stone as shown here is easier to keep at a constant bevel than one ground with a separate honing angle.

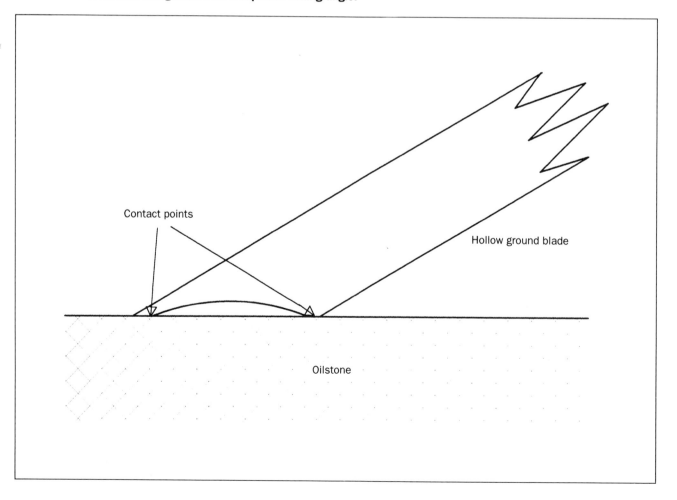

Contact points

Hollow ground blade

Oilstone

grinder of some sort. (Grinders and grinding are discussed in Chapter Three.)

Many modern grinding machines have small diameter wheels. These do not produce a flat surface, but grind the surface with a hollow, hence the term **hollow ground**. This has led to a different sharpening technique: instead of just honing the very edge of the blade, the two extremities of the hollow are presented to the stone. This keeps the blade at a set angle on the stone (see Fig 2.3). It is part of the art of sharpening to keep the tool at one angle when honing. This is quite a difficult skill to acquire. The hollow ground technique, while not doing away with the need for this skill, simplifies it.

THE SINGLE BEVEL

The Stanley Bailey planes and all their copies have thin blades. One of the original selling points was a statement that the blade never needed grinding. Because of the thin blade, the honing surface never became wide. However, most craftsmen that I know still use a double bevel on these blades. Where a craftsman continuously works away from the workshop, and does not have access to a grinding machine, this **single bevel** might be an advantage. I have met one or two craftsmen who sharpen all their tools with just the one bevel. They claim that although this takes longer to sharpen they get a better edge.

THE MICRO BEVEL

There is a further sharpening technique that has quite a strong following, particularly in recent years. A tool is sharpened in the normal way and then it is drawn back along the stone a couple of times at a steep angle (see Fig 2.4). Advocates of this method of sharpening say that the **micro bevel**, being at a steep angle, lasts longer than a conventional edge. The fact that the micro bevel is very small means it does not hinder the tool entering the wood. I have seen a micro bevel

applied to both faces of a cutting edge. This flies in the face of all that I have been taught about sharpening, but it seems to work for some people.

When a strop is used to put the final finish on an edge, it tends to cause a slight rounding of the bevel. This, I suppose, could be likened to a micro bevel, however the effect is infinitesimal. The strop is a very useful item, particularly when carving or doing some delicate paring job where a really keen edge is required. From the moment we start to use a freshly sharpened tool it begins to lose its edge but, by frequent stropping, a very sharp edge can be maintained. The strop is kept on the bench close to hand and every few minutes the tool is stroked over its surface a couple of times. (Strops and stropping are dealt with in Chapter Six.)

GETTING ORGANIZED

As sharpening occupies a fair amount of non-productive time there is a need to keep it within reason. This is best achieved by keeping the edge of the tools in good condition. Regular honing only takes a few moments, whereas an edge that is in poor condition can take as much as 30 minutes to put right. It is clear, therefore, that putting off sharpening a blunt tool does not save time. Not only is there the sharpening time, there is the extra time it takes working with a tool that is out of condition,as well as the frustration caused to the user.

When I was earning my living at the bench, I allowed four hours a week non-productive time. This was for sharpening, washing hands and keeping time and material records. Most of the four hours were spent sharpening. Nearly all the work was carried out to a very high standard, using hand tools, and this required that the tools were always in prime condition. There are several things that can be done to reduce the time spent sharpening.

Make a small area in the workshop a sharpening station and place here, close to hand, everything that is needed for the sharpening operation. The way you lay out the equipment will of course depend on what system you adopt. Sharpening stones can be kept laid out on a surface at the correct height for immediate use. When they are not in use, they should be covered to prevent contamination by the ever-present workshop dust. Don't keep the stones in a drawer. Every time you need them the drawer has to be opened, the stones taken out, used, wiped and put away again. I have yet to find the craftsman who enjoys sharpening. The easier it is made the less of a chore it will become. The sharpening

station needs to be near the bench so that as little time as possible is spent in walking to it.

KEEP IT CLEAN

The sharpening process generates dirty by-products, such as black oil and metal particles. If these are allowed near woodwork they will contaminate it. It is important to keep the sharpening station as clean as possible. The dust produced by high-speed bench grinders is not only gritty and dirty, it is a health hazard, and industrial premises that come under the auspices of the

Fig 2.4 Micro bevel cutting edge.

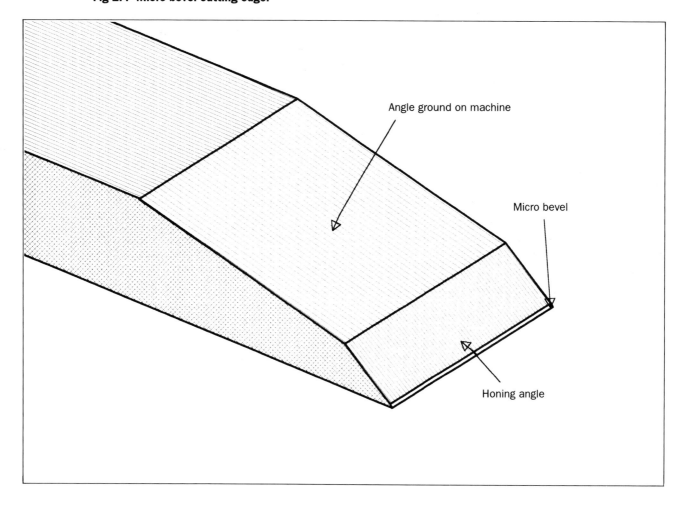

Angle ground on machine

Micro bevel

Honing angle

factory inspectorate have to be fitted with dust extraction. This is quite a worthwhile investment for any craftsman who is concerned with his future health. An old disused cylinder vacuum cleaner is quite adequate for this purpose. It will be found that most dry grinding machines have an outlet at the back of the wheel guard (wet grinders do not make dust). The size of this outlet is nearly the same as the normal domestic vacuum cleaner pipe. The electric supply to the cleaner can be rigged through the grinder switch. In this way the dust extraction will come on when the grinder is switched on. Loose abrasive and metal particles are retained by the cooling liquid.

Some craftsmen have a habit of picking up a handful of shavings and wiping the oil from the previously sharpened tool. This is a precarious practice. It is so easy to get oil on the hands and thus onto the work. Not only that, but what happens to the contaminated shavings? Are they returned to the workshop floor where they become a hazard?

Sharpening needs to be carried out in a good light. The sharpening station should not be relegated to a dingy corner of the shop. I have a small strip light over my stones, turned on by a conveniently placed switch. Thus the light is only on when I need it. Saw sharpening (*see* Chapter Eleven) is best carried out in natural light. As this is not such a regular item as honing it can usually be done when there is good daylight. The process is best carried out in front of a north-facing window, in the early afternoon. It is not a process that should be interrupted and the time picked to undertake the task needs to be chosen with some care. However, most craftsmen have to fit in saw sharpening between jobs, and the ideal time and place are not always possible to find.

A Sobering Thought

Every time we sharpen a tool we remove metal from it, and a little of its life has gone. This, if we think about it too hard, can leave us in a dilemma. Here we have a very fine tool – let us assume it is a chisel made many years ago from the best cast steel. This is a favourite tool; it is picked up and used in preference to any other chisel. I think every craftsman has a chisel that he tends to favour. However, as we subject this tool to a lot of use, it follows that it is being frequently sharpened, and perhaps there is an argument here to share the work out more fairly among the tools. Saving that very special tool for special tasks will prolong its life. Most craftsmen who spend a life in the trade wear out at least one set of chisels. These shortened chisels still have their uses. Quite often they are the only tools that can be used in a confined space. Apprentices usually scrounge these old chisels from a retiring craftsman. I have known them to become more prized than a new, full-length chisel. Grinding, in particular, reduces the life of a tool and should not be overdone. When grinding, leave a honing surface at least $\frac{3}{64}$in (1mm) wide. However, as grinding saves time, don't let this honing surface become wider than about $\frac{1}{8}$in (3mm). Never grind the honing surface right away. This is really wasting metal.

CHAPTER 3

GRINDING

WHY GRIND?

Rubbing a blade back and forth on a stone is a laborious task. For centuries this task has been lightened by using a **grindstone**. The grindstone consists of a fine-grained sandstone wheel, usually about 27in (686mm) in diameter, and 4 or 5in (100 or 125mm) thick. This is mounted on a stout stand with its lower half encased in a trough containing water. The wheel is turned with a cranked handle, or sometimes by a treadle. Some grindstones are equipped with a pair of metal arms into which the tool to be ground can be clamped. This allows the operator to exert considerable pressure on the tool. The force pushing the tool against the stone can be so great that it becomes difficult to turn the crank.

This turning job, in days gone by, was undertaken by some menial, such as the apprentice, while the craftsman who owned the tool presented the tool to the stone. In my early days in the trade it was considered very funny by some men to suddenly release the weight they were applying to the tool. The crank would then fly round, causing the person turning it to fall on the ground. These old grindstones that run in water and grind at low speed make a very good job of grinding the edge of woodworking tools, but unfortunately the task is still somewhat arduous. With the arrival of electricity in most workshops the grindstone had a motor attached to it, and before long smaller and faster machines were installed. However, where time and speed are unimportant, there is much to recommend these old-fashioned machines. The copious supply of water that cools the tool, the size of the stone, and the speed with which it cuts, all add to the appeal for many craftsmen.

Unfortunately, most surviving machines have been spoilt because water has been left in the trough. This softens the part of the stone that for long periods remains below the surface, so that when used the soft part wears away much faster than the rest of the stone. Thus the stone rapidly loses its shape, and is of no further use.

THE HISTORY OF GRINDING

The grindstone in its basic form is described and illustrated in the *Utrecht Psalter* that was published around 850 AD. This is the earliest reference I have been able to find. Israel von Meckenhem drew and described a treadle-operated grindstone in AD 1485, and around 1500 Leonardo da Vinci designed several grinding machines. These were all power-driven, either by horses, wind or water. By 1570, powered grindstones were quite common. These were not used by woodworkers but by the military for sharpening their weapons. The requirements of the army with their complicated armour, which needed constant polishing, and weaponry such as swords, pikestaffs and spears, led to the installation of some large-diameter, power-driven stones. It seems that man has always applied his mind to improving the weaponry of war, in preference to the more productive articles of peace.

All these machines used a wheel made from natural, fine-grained sandstone. Meanwhile, China and other Far Eastern countries were making grindstones from natural emery. This was fused into a wheel using lac, a resinous substance secreted by certain coccid insects,

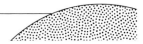

but these wheels were not strong enough to withstand high-speed revolutions. It was the Industrial Revolution in the nineteenth century which led to the development of a man-made abrasive material for the engineering industry. Until the end of the Second World War woodworkers mainly used the old-fashioned sandstone wheel, more often than not turned by hand. Some cabinetmaking firms in the east of London had grinding arrangements with tool shops who offered this service. The craftsmen had 6d (2½p) per week grinding money deducted from their wages to pay for this service.

SIZE OF STONE

Most grinders are used so that the edge of the wheel does the work. That is to say, the tool being ground is presented to the periphery of the stone. The diameter of the wheel affects the face of the ground surface, so that the smaller the wheel the more pronounced is the hollowness of the ground surface. Some people like a flat ground surface, and for them a machine designed so that the tool can be presented to the side of the wheel is required.

One word of warning here: when grinding on the side of the stone, a special wheel is needed.

Fig 3.1 Grinding an in-cannel gouge on a gouge cone.

Under no circumstances should a wheel intended for use on its edge be used. The stone is not designed to withstand the forces imposed on its side by this method of use. Beware! A grindstone breaking up at 3,000rpm can cause considerable damage.

The speed at which the spindle of the machine revolves affects the size of the stone. Modern high-speed grinders run at 5,000 to 6,000 feet per minute. Most new wheels have the maximum speed at which they can be safely used

Fig 3.2 The ideal safety gear for use when truing a grindstone. Fresh air is drawn in through a filter at the top and blown down over the face; stale air is exhausted at the bottom.

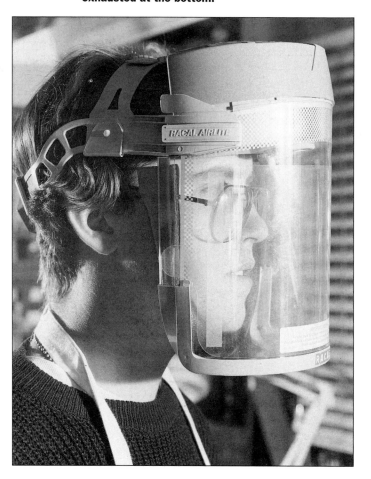

printed on the cardboard washer attached to their side. Occasionally a requirement arises making it necessary for a new wheel to be installed on a machine. Because of constant use, and frequent dressing, the size of the existing wheel has been reduced, and it is impossible to discover its original diameter. Some simple arithmetic will establish the diameter in inches of the replacement wheel required. The surface speed in feet per second is multiplied by 12 to change it to inches. This is divided by 3.142 multiplied by the spindle speed in revolutions per minute. Most of the high-speed bench grinders offered by retail tool shops have 6in (152mm) wheels.

When it comes to grinding shaped cutters for moulding machines or moulding planes, a narrow wheel is required. This will quite often have a rounded edge so that it can follow the required contour of the cutter's edge. In wood mills, where there is a need to produce various mouldings, several double-ended grinding machines are installed. Each wheel fitted to these machines has a different edge contour. There is also a stone that can be found on some grinders called a **gouge cone** (*see* Fig 3.1), which is used to grind the bevel on an in-cannel gouge.

SAFETY

The grinding machine looks quite harmless, particularly in a workshop where there are woodcutting machines. Everybody is aware of the fast-revolving cutters, or saw blades. By contrast a well-balanced stone running sweetly appears not to present any hazard. How wrong! I have previously said that the surface speed of a bench grinder is 5,000 feet per minute, equivalent to 56.82mph (91.44kph) The wheel is made of abrasive grains held together with a vitrified material. This is quite a fragile combination. Should the wheel break, pieces of stone will be thrown about the workshop at speeds in excess of 50mph (80kph). This is why particular attention should be paid to mounting the stone properly (*see* page 22). When the machine is being used, minute pieces of the material the wheel is made from are constantly being flung into the air. There is also the metal that is being abraded from the tool, most of which leaves the surface in a shower of sparks. Eye protection must be worn at all times when grinding on a dry wheel. FORESIGHT IS BETTER THAN NO SIGHT. The spark guard on the grinder is there to stop hot sparks, not as eye protection. It is no substitute for safety glasses (*see* Fig 3.2).

The grinding wheels on all modern high-speed grinders are partly enclosed. This enclosure guards about two-thirds of the wheel's circumference. Never take the guard off to use the side of the wheel. At the back of the guard, which is the part furthest away from the tool rest, there is a spout sticking out for the attachment of a dust extractor. An extractor is required by law under the Health and Safety at Work Act. While this regulation does not apply to the home workshop, some form of extraction should be considered. I discussed adapting a domestic vacuum cleaner for this purpose in Chapter Two. Even when extraction is installed, a face mask should be worn when dressing the wheel, as a lot of abrasive dust is released into the atmosphere.

Last but not least, remember if the wheel is capable of removing metal it will certainly remove flesh at an alarming rate. Only a slight brush of the hand against the revolving wheel causes a nasty abrasion. Be careful.

Fig 3.3 Ring testing a grinding wheel.

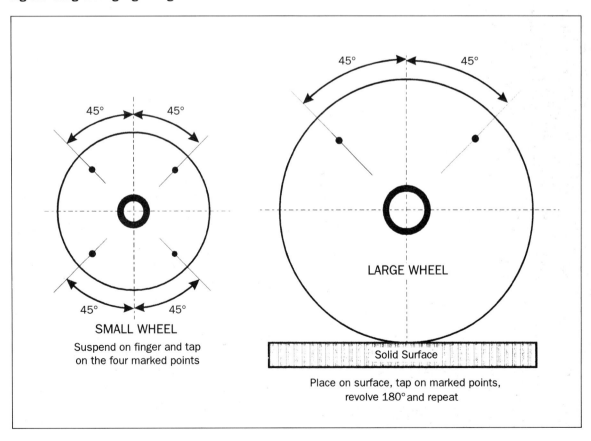

45° 45°

45° 45°

SMALL WHEEL

Suspend on finger and tap
on the four marked points

45° 45°

LARGE WHEEL

Solid Surface

Place on surface, tap on marked points,
revolve 180° and repeat

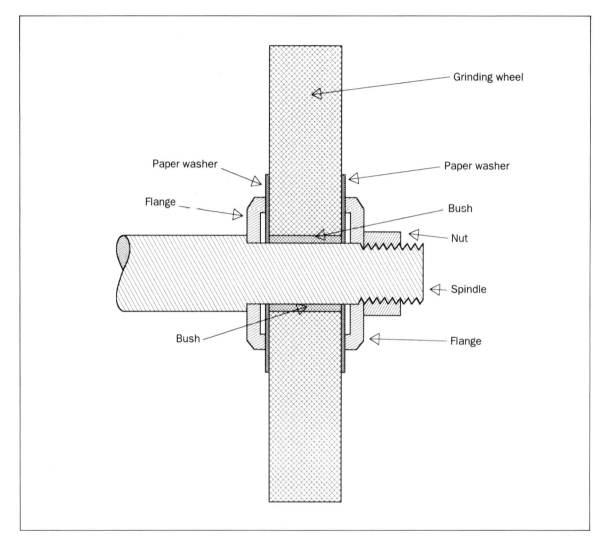

Fig 3.4 Section through a correctly mounted grinding wheel.

Mounting a Wheel

At the centre of the wheel there is a lead or plastic bush. If the bush is plastic, and the hole at its centre is the wrong size, it is possible to change it for a bush with the correct size hole. The hole in the bush should be a good fit on the machine spindle. Before mounting the wheel, inspect it carefully for cracks. Next, a **ring test** should be carried out. Put your little finger through the hole at the wheel's centre, so that it is suspended freely. Now with a non-metallic item, such as a chisel or screwdriver handle, tap the wheel as shown in Fig 3.3. Undamaged wheels give off a clear ringing tone. A dull tone means a cracked wheel. Never use a cracked wheel. Large-diameter wheels can be stood on a surface to ring test them. They need to be turned 180° after the two top quadrants have been checked.

The driving spindle will have two relieved flanges. These are relieved so that they only grip the wheel with their outer edges. The inner flange is usually fixed to the spindle while the outer one is free. Paper or thin card washers are placed either side of the wheel between it and the flange. The flange should be at least a third of the wheel's diameter. Fig 3.4 shows a section through the assembled wheel, paper washers, and flanges. The nut that secures the assembly onto the spindle should not be over-tightened. This can cause damage to both the wheel and the flanges.

DRESSING THE WHEEL

Once the wheel has been mounted, it should be **dressed** in order to balance it. This is not the only time when the wheel needs dressing. From time to time the surface of the wheel becomes clogged up with material that has been ground. The cutting surface also becomes dull and needs removing so that another layer of abrasive is revealed. Dressing is carried out by using a tool which is presented to the cutting face of the wheel. There are basically three types of tool. The first (*see* Fig 3.5) consists of several star washers mounted side by side in the end of a metal handle. The washers are pressed against the wheel, and as they revolve, the tips of the stars knock small particles from the grinding surface. Secondly a stick of silicon carbide stone can be used to abrade the grinding surface. Sometimes this stick is enclosed in a plastic handle. A single point diamond mounted in the end of a metal bar is also a very effective tool, but somewhat expensive to buy. Dressing the wheel removes the main cause of vibration, and once the wheel is running true there should be very little vibration. Sometimes the side of a new wheel has to be dressed to reduce out-of-balance forces. This

Fig 3.5 Star wheel grindstone dresser.

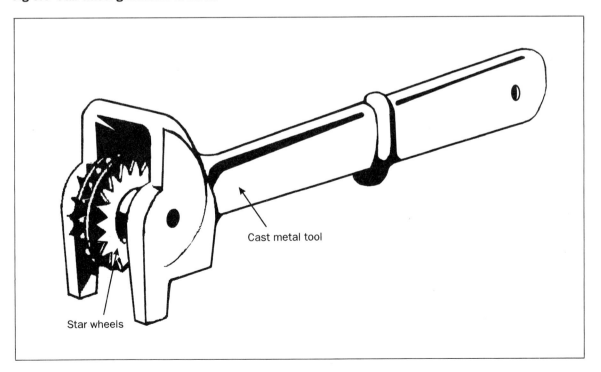

Cast metal tool

Star wheels

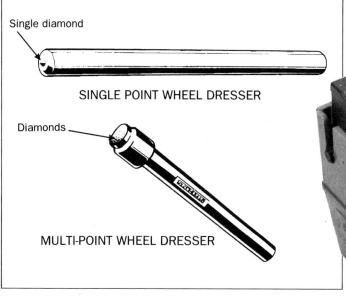

SINGLE POINT WHEEL DRESSER

Single diamond

Diamonds

MULTI-POINT WHEEL DRESSER

Fig 3.6 Diamond-tipped wheel dressers.

Fig 3.7 A small wet grinder: an ideal machine for workshops where space is limited.

needs to be done with care; do not apply too much pressure to the side of the wheel. Fig 3.6 shows two examples of diamond-tipped wheel dressers.

DIFFERENT TYPES OF GRINDING MACHINE

There are many different grinding machines available, and the choice that one makes will depend on just what it is to be used for. Some woodturners use most of their tools straight from the grinder; they do not hone the edges at all. A carver, however, hardly ever grinds a tool. There is also the matter of personal preference. Is one prepared to take longer over the task of grinding a tool on a fine-grained, water-cooled stone, or is speed important? If the latter, then a high-speed bench grinder should be chosen. There are many models and makes of high-speed grinder, but when it comes to water-cooled machines the choice is somewhat limited. It should also be remembered that it is very easy to burn a tool

using a fast, dry stone. The term **burn** is used in the trade to describe the overheating of a tool. The tool's steel when heated beyond a certain temperature becomes soft and will not keep a sharp edge. When this happens, it is said to have **lost its temper**.

Another matter of preference is the choice of flat or hollow grinding. One also has to consider how much of the workshop space the machine can be allowed to occupy (*see* Fig 3.7). After all is said and done, while sharpening is important it is not the primary purpose of the workshop, and there are many pros and cons to be weighed up before rushing out and buying a grinder.

Water is not the only cooling agent used. I have been employed in workshops where the grinder has been cooled with oil. The best-known machine that uses oil is probably the Viceroy Sharpedge made by Denford. This machine is a bit on the large size for the home workshop, and is more likely to be seen in a professional workshop where several craftsmen would use it . The Viceroy Sharpedge is also the grinder found in most school workshops; you may have seen one during your schooldays. The machine has a flat grinding wheel which looks like a record turntable. A spray bar keeps a constant film of thin oil on its surface, and there is a jig to hold the tool being ground at any pre-set angle. Like all machines that rely on a liquid coolant, the speed of the grindstone is limited. Should the wheel turn too fast the coolant would be thrown all over the workshop by centrifugal force.

Another design of oil-cooled machine has a wheel that is hollow at the centre. This hollow centre forms a sump that is filled with oil. As the grinding wheel revolves, the oil is forced through the porous stone to its outer surface where the grinding takes place. This machine tends to be very dirty in use, since oil escaping from the grinding surface picks up all the grinding dust. The whole gooey mess collects on the surface of the tool being ground.

All liquid-cooled stones are limited in their speed by the need not to throw the liquid about. However, the owner of quality tools should not be impatient when grinding them. It is far better to use a cooled machine than to spoil a prized tool by over-heating.

A machine that has become very popular with woodworkers was originally designed to sharpen butchers' knives. The Sharpenset now sells to more woodworkers than butchers, and the manufacturers have produced several attachments to make the machine more useful in the woodwork shop. There are attachments for holding chisels and plane irons at predetermined angles while they are being ground, and a planer blade sharpening jig is also available. This machine is very compact and occupies a minimum of space; I have owned one for over 20 years and it is still giving sterling service. The only fault I have found is the corrosion which afflicts the shaft on which the stone is fixed, because it is always wet. This makes changing grinding wheels a chore. Also the grinding surface offered by this machine is a bit on the small side.

THE NOT-SO-GOOD MACHINES

There is hardly anything in this world that somebody won't make a little worse and sell more cheaply. This is certainly true of grinding machines. I recently saw a brand new copy of a Japanese machine. This machine, made in Taiwan, had to be seen to be believed. In appearance it was the same as the Japanese machine, however, the difference became apparent the minute it was turned on. The vibration was unbelievable. Even after both stones had been trued there was still serious vibration. On close inspection most of the machine was seen to be made of thin plastic, allowing the parts of the machine to flex and making the tool rest ineffective. This design of machine has a reservoir of water that is allowed to trickle on to the surface of the flat wet stone. Water then collects in a receptacle below the machine, which when full is emptied into the top reservoir. The original Japanese machine is ideal for the fine woodworker, but don't buy the Taiwanese copy (*see* Fig 3.8).

KEEPING THE TOOL COOL

If you are fortunate enough to have the space and money to install two grinding machines, you can have the best of both worlds: a cooled machine for grinding the edge of prized cutting tools, and a high-speed dry grinder for removing metal quickly. However, provided one has a gentle touch, it is quite possible to grind all tools on a high-speed machine without spoiling them. The secret is the amount of pressure used to push the tool against the stone. Some workers use one hand to hold and guide the tool being ground, while they spray a fine mist of water over the tool's edge (*see* Fig

Fig 3.8 A horizontal wet wheel and high-speed grinder from Taiwan. The wet stone is nearer to a hone than a grinding wheel. The machine has severe vibration even after the stones have been trued.

3.9). Although this keeps the tool cool, it introduces moisture into the atmosphere, which is not always acceptable in the woodwork shop. The method generally used to cool a tool being ground is to dip it frequently into water. Some high-speed grinding machines have a receptacle built into the case for this purpose, but most craftsmen make do with an old tin can.

I have used a can of water when grinding, but I must admit that I get carried away and forget to dip the tool into it. This only needs to happen once, and the tool is spoilt. For this reason I use

two grinding machines. The only tools that I now have to dip into water to cool are in-cannel gouges. This is because my gouge cone fits on the high-speed grinder. As the circumference of the cone is small, the grinding speed is much lower than that from the normal grinding wheel.

If you are using a grinder for the first time, experiment with something that does not matter. It is very instructive to grind several pieces of bright steel and see the colour change. The thickness of the metal affects this; a piece of thick steel bar will absorb far more heat than a thin bar before colour changes occur. This explains why it is so easy to burn the very thin part of the tool near the cutting edge.

The different colours that show on the surface of bright steel as its temperature rises are an indication of just what is happening. As the temperature at the grinding surface rises and dissipates up the blade a whole range of colours can be seen running up the tool. These include a very pale straw colour (appearing at 430°F), brown-yellow (500°F), and dark blue (570°F). These temperatures may seem high, but you will be surprised how easy it is to obtain them when grinding. Any discolouring means that the characteristics of the steel have been changed.

Because it is the temperature that causes the

Fig 3.9 Water being sprayed while grinding on a high-speed grinder.

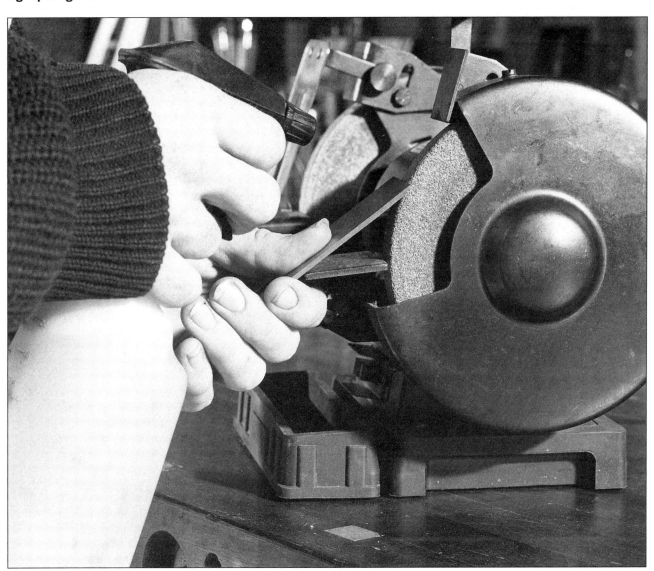

change, and this is indicated by the colour, one can see what has happened, and it is often possible to grind away the coloured part of the tool. This restores it to the original condition, but of course much of the tool's life has been thrown away in grinding dust.

INSTALLATION

Unless you are investing in a large industrial machine, the grinder will need mounting on a stand or bench of some sort (*see* Fig 3.10). It is best to make a special stand that puts the grinding wheels at the ideal working height. This ensures that you will have the proper control over the tool being ground. A special fitment can be made that has storage space for all the items

used when grinding, including grinding wheels that are not fitted on the machine. Some machine manufacturers supply a pedestal as an accessory. I have used grinders mounted on one of these, and provided they are fixed to a solid floor they are fine. If they are mounted on a flexible floor the machine feels unstable, as it moves slightly under working pressure. I prefer a wooden structure with room around the grinder where I can put items that I need to use. I have a small engineers' try square, and a bevel template that I like to have close to hand when I am grinding a tool.

THE GRINDING WHEEL

There is a lot more to a grinding wheel than you might think. It is really thousands of small cutting

Fig 3.10 Sections through alternative methods of fixing grinder to bench.

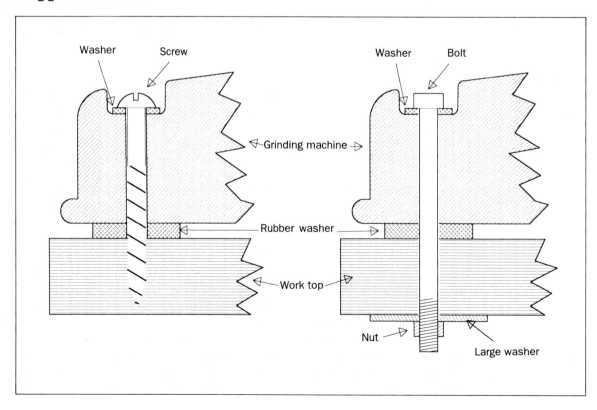

tools assembled into one larger tool. These small tools are grains of abrasive material held in a bonding agent. There are many different types of wheel, varying in terms of their abrasive material, the size of the grains, the bonding agent, and the density of the grains.

In use, the grains of abrasive material are constantly breaking up. Every time a grain fractures or breaks away from the surface of the wheel, a new minute but sharp cutting edge is produced. The selection of the correct grinding wheel is very important if the grinding process is to be carried out efficiently.

There are four main types of abrasive grain: **aluminium oxide**, **cubic boron nitride**, **diamond**, and **silicon carbide**. Cubic boron nitride and diamond are unlikely to be used in the woodwork shop, but the other two have properties that are of interest to us. Aluminium oxide is a hard crystalline substance made by heating bauxite in an electric arc furnace. Silicon carbide is an artificial abrasive made by heating a mixture of sand, coke, sawdust and salt in a resistance furnace. Manufacturers give these materials trade names which can be confusing. Silicon carbide is marketed as Carborundum and Crystolon by different companies. Aluminium oxide is sold under the names of Aloxite and Alundum.

For most purposes you will require aluminium oxide. Silicon carbide will only be used to grind tungsten-tipped tools such as masonry drills, router bits, etc. Some planer blades have tungsten edges, and it is here where the hard-cutting grains of silicon carbide are useful.

The abrasive grains that comprise the cutting medium in the grinding wheel can be packed into the bonding material at different densities. That is to say they can be packed either close together or spaced apart. This is achieved by using a filling agent during the wheel's manufacture. When the wheel is vitrified, the filling agent is burnt away leaving openings evenly distributed throughout the wheel. The more filling agent used the less dense the wheel. Added to this there are five main

Very soft	Soft to medium	Medium to Hard	Hard	Very Hard
A	E	M	U	V
B	F	N		W
C	G	O		X
D	H	P		Y
	I	Q		Z
	J	R		
	K	S		
	L	T		

Fig 3.11 Table showing relationship of alphabetical classification to wheel hardness.

bonding agents. These are vitrified ceramic, phenolic resin, rubber, shellac, and metal. All these variables can be used in any combination, so you can see there are many permutations.

There is yet another variable, known as the **hardness** of the wheel. This is determined by how strongly the bonding agent holds the grains of abrasive. With a soft stone, grains are easily removed from the surface of the wheel as it is being used, and new cutting edges are being produced all the time. Conversely, in a hard stone the grains are held very securely. The hardness is determined by the type and amount of bonding material used.

The degree of hardness is stipulated by a code letter ranging from A to Z, 'A' being very soft and 'Z' extremely hard (*see* Fig 3.11). Stones fitted to most grinders available through the retail

**Fig 3.12
Grinding
wheel codes.**

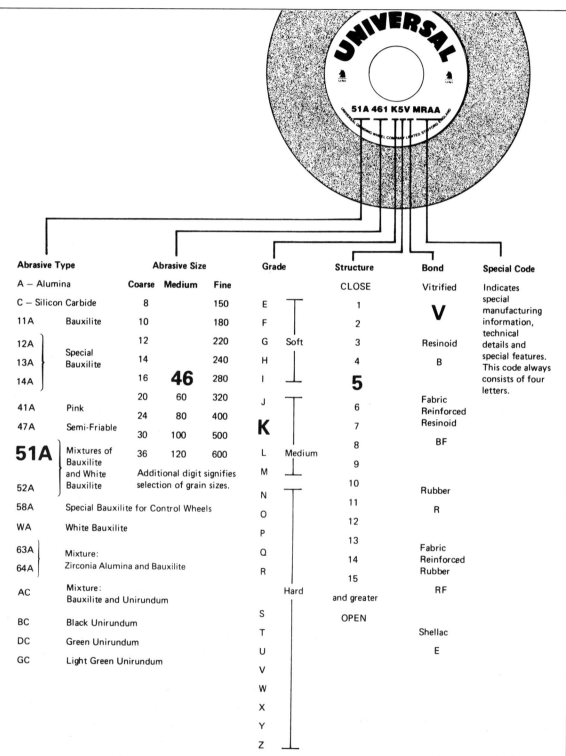

Abrasive Type		Abrasive Size			Grade		Structure	Bond	Special Code
A — Alumina		**Coarse**	**Medium**	**Fine**			CLOSE	Vitrified	Indicates special manufacturing information, technical details and special features. This code always consists of four letters.
C — Silicon Carbide		8		150	E		1	**V**	
11A	Bauxilite	10		180	F		2		
12A	Special Bauxilite	12		220	G	Soft	3	Resinoid	
13A		14		240	H		4	B	
14A		16	**46**	280	I		**5**		
		20	60	320	J		6	Fabric Reinforced Resinoid	
41A	Pink	24	80	400			7		
47A	Semi-Friable	30	100	500	**K**			BF	
51A	Mixtures of Bauxilite and White Bauxilite	36	120	600	L	Medium	8		
					M		9		
52A		Additional digit signifies selection of grain sizes.					10	Rubber	
					N		11	R	
58A	Special Bauxilite for Control Wheels				O		12		
WA	White Bauxilite				P		13	Fabric Reinforced Rubber	
63A	Mixture: Zirconia Alumina and Bauxilite				Q		14		
64A					R		15		
AC	Mixture: Bauxilite and Unirundum					Hard	and greater	RF	
BC	Black Unirundum				S		OPEN		
DC	Green Unirundum				T			Shellac	
GC	Light Green Unirundum				U			E	
					V				
					W				
					X				
					Y				
					Z				

trade are around 'N'. Soft wheels help prevent the material from overheating, but wear away quickly. Hard stones tend to glaze and need frequent dressing.

Grinding wheels usually have thick paper washers fixed to their sides. One of these washers is usually printed with a code. This code is the wheel's specification. If you have a new machine that has been delivered with wheels attached you may have to remove the nut and flange to read this code. The specification is a group of letters and numbers as shown in Fig 3.12.

WHAT DOES ONE REALLY NEED?

All this technical information will enable you to find out what you have. This will not necessarily be what is ideal for grinding woodworking tools. Remember we are talking here about high-speed bench grinders, which are used dry. For most woodworking tools made of high-carbon steel and hardened between Rockwell C 58° and 62°, the following would be ideal: aluminium oxide (grey), of hardness between grades N and P. Two wheels are required, one medium about 36 grit and the other a fine 100 grit. A couple of numbers either side of these would not make much difference (see Fig 3.13).

In most workshops there are tools made from high-speed steel, particularly turning tools. For these tools a soft (F-K) aluminium oxide (pink or white) wheel is more suitable than those I have suggested for high-carbon steel. Tungsten carbide is used to tip many machine tool cutters. This cannot be ground with aluminium oxide. A silicon carbide (green) wheel is required. An aluminium oxide wheel applied to tungsten will just glaze over and probably heat up the tool you are trying to grind. This heat can (particularly on small tools) melt the brazing used to hold the tungsten on to the tool. Machines that grind while applying a cooling agent to the surface of the wheel are

Coarse	Medium	Fine	Very Fine
12	30	70	150
14	36	80	180
16	46	90	200
20	60	100	240
24		120	

Fig 3.13 Table showing numerical classification of grit size in grinding wheels.

supplied with a wheel that is suitable for the coolant.

Nearly all the wet grinders are supplied with some form of tool holder that will present the tool to the wheel at a predetermined angle. High-speed bench grinders normally only have a tool rest. This can lead to multiple bevels being applied while grinding. It is quite a simple task to make wooden attachments that can be put on top of the tool rest. These are made so that the tool being ground rests flat on their upper surface, while its edge is presented to the wheel at the required angle. Several blocks can be made, each with a different angle. Re-grinding a tool is then just a matter of using the correct block (see Fig 3.14).

A LIGHT HAND

I have described the various grinding machines and wheels at considerable length. We now come to the practical part of actually grinding a tool. The technique is slightly different depending on the machine and whether there is a jig to hold the tool. On all machines it is normal to work with the wheel turning towards the tool being ground.

The first consideration is the angle at which the edge is to be ground. Some form of rest or jig

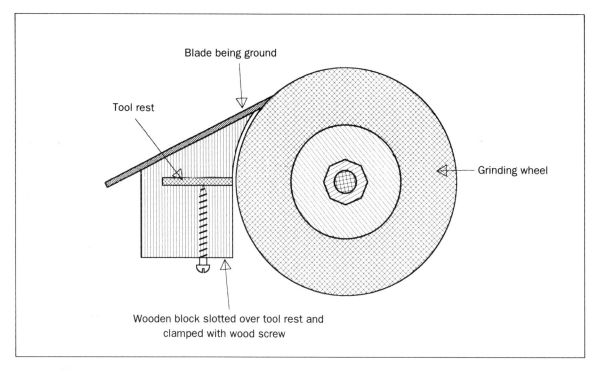

Blade being ground

Tool rest

Grinding wheel

Wooden block slotted over tool rest and
clamped with wood screw

**Fig 3.14 Block fitted to grinder tool rest so that
tool is ground at the correct bevel.**

must be used if the task is to be carried out with
any precision. Once the tool is placed on the rest
or in the jig it is pressed against the surface of the
wheel. The pressure used should not be excessive.
It is surprising how little pressure is really required,
particularly on a high-speed bench grinder. The
novice might think that the more pressure applied,
the more quickly the task will be completed. This is
not so; all that happens is that the heat caused by
friction quickly builds up and the tool is overheated
and spoilt. Undue pressure does not allow the
grains of abrasive to cut properly. The actual task
of grinding requires a light, delicate touch, and a
smattering of common sense.

Once the tool is in contact with the wheel it
should be slowly moved from side to side. On a
wheel where the grinding is done by its periphery,
this movement ensures that the whole width of
the tool's edge is ground evenly. Where the
grinding is carried out on the flat side of a wheel,

the movement across the radius wears the
surface of the wheel evenly. A most important
feature of grinding is to keep the tool cool. When
dry grinding, the frequent submersion of the tool in
cold water is essential. If you hold the tool on a
rest the heat can be felt with the fingers. As
previously stated, it is a good idea to experiment
with an old tool or piece of steel to get the feel of
the task. In doing this one can purposely burn the
test piece, and see just how and when it happens.

Frequent grinding can reduce the time taken
up sharpening the tools. Under no circumstance
can a tool be termed sharp when it leaves the
grinder. Do not grind a tool right to its cutting
edge, leave a small flat at the honing bevel.
Some scraping tools used for woodturning are
sharpened at an extremely acute angle, and used
straight from the grinder. These tools are used to
scrape not cut, and their edge cannot really be
described as sharp.

CHAPTER 4

HONING

There are many different ways of honing. I have closely observed the process whenever I have had the opportunity. There may not be much difference between the technique used by one craftsman and another, but differences there are. I will describe the different stones, lubricants, guides and methods, and from these you can choose what is suitable for your own purpose.

Experiment until you find the best combination. If you are in the fortunate position of being able to try honing on a variety of stones, that is fine, but the person working on their own with only the equipment they buy will be somewhat limited in their experimentation. I will endeavour to describe every type of stone and device that I have come across, and from this description you can make an informed decision on what will suit you best. You may already have items that, with a little work put in to return them to top condition, will serve you well.

IT'S ALL IN THE STONE

The grindstone removes metal quickly, but it gives a rough, coarse edge. Something is needed on which to rub the tool which, while abrasive, is sufficiently fine to impart a smooth finish resulting in a keen edge.

A fine stone on which to hone your tools is a thing to be desired. There are many medium and coarse-grade stones that will almost put the edge on a tool. It is that final polishing which requires just the right stone. When talking about the right stone I am reminded of the monologue that Sir Bernard Miles used to relate, when he played the part of a country yokel. It describes the lengths to which craftsmen will go in search of sharpness:

Our old vicar was showing some tourist round the church the other day. They came to a tomb with an effigy of an old crusader on it, he's got a little crusader dog at his feet, but all his face is missing. The vicar says to the tourists, this 'er Crusader was defeated by Holiver the Cromwell. Well, I 'ad to laugh 'cause 'e don't know as how we creeps into the church at night and sharp's our faggin hooks on 'im. He's the best bit of sharpening stone this side of 'Artfordshire.

I will not go as far as to suggest that you go around surreptitiously rubbing your pen knife on statues and the like to test their sharpening qualities, but keep your eyes open and see what others are using. Stones fall into two distinct categories: natural and man-made. Man-made stones tend to have a uniform structure with no variation through the stone. Much natural stone is rejected to find a piece that will make a reasonable-sized, uniform sharpening stone, and consequently, some natural stones command a very high price. Certain once highly prized stones have all been quarried, and the only source is the secondhand market.

Old craftsmen relate stories of their youth when they saw stones that other craftsmen used. These stones, so they said, possessed almost magical powers, and could impart a cutting edge that had to be used to be believed. As an apprentice I would listen with interest while the craftsmen discussed sharpening. It always surprised me how each considered his method superior to those of his workmates. There were always stories of some old craftsman who used a secret formula to mix the lubricant he used on his stones. This again was so superior as to be almost legendary. Over the years I have spent a small fortune on stones, many of which have only been used a few times before being discarded for something better. I have yet to find the stone with the magic in it, though I think I may have come near it once or twice. In my opinion, Japanese water stones, both man-made and natural, are the ultimate. (Japanese tools are discussed in Chapter Sixteen.)

NATURAL STONES

When you consider that the planet earth is composed mostly of rock, you would think that natural sharpening stones ought to be easy to find. However, there are surprisingly few places in the world where quality stone suitable for sharpening exists. The best-known contemporary location in the Western world is probably Hot Springs in the Ouchita Mountains in the USA, where novaculite is quarried. This is the famous **Arkansas** stone. As you would expect with a naturally occurring material, no two pieces are quite the same. There are three distinct Arkansas stones: hard black, hard white and soft white. There was once a translucent, very hard white variety. Unfortunately this superb stone is now only available in small pieces. The main deposits have all been quarried. The black is the nearest thing obtainable to the translucent white. This is a fine-grained stone that imparts a very sharp edge. The soft white cuts much faster than the black and is of a coarser grained structure. Some Arkansas stones are pied, that is to say they are composed of both black and hard white stone. One might think that the pied stone would wear unevenly, but I have not found this to be so. **Washita** stone is a second grade of Arkansas. It comes from the area around the Ouchita River. This stone is softer and of a coarser structure than the best Arkansas. It makes a good intermediate stone.

In the UK the Tam o' Shanter, Charnley Forest, Rag, Water of Ayr, and Welsh have all been quarried and marketed. I have a small Tam o' Shanter stone which is very soft and fine-grained.

It wears fast compared with my other stones, but it gives quite a fine edge. The only other British stone that I have much experience of is the Welsh. This is a slate and, though on the soft side, it has a fine-grained structure. This stone is obtainable in large sizes, and is reasonably priced. It too wears fairly fast and needs frequent truing. This is not an arduous task; being soft, the stone can quickly be dressed.

Charnley Forest is a green-coloured stone of slate character. One or two older craftsmen that I have worked with have used this stone. Its fine grain should make it a good finishing stone. Those that I have used I found very slow cutting. Charnley Forest is also known as Whittle Hill Stone. Water of Ayr, (also known as Scotch stone or Snake stone) is too soft for honing general woodworking tools.

I was once allowed to use a piece of Greenstone from Snowdon. This was like slate and also very fine-grained. It was once prized for sharpening surgical instruments. The woodworker able to obtain a piece of this stone would be considered lucky. It is extremely fine-grained, and cuts very slowly.

Over the years, several varieties of stone quarried in Europe have been successful in this country. Turkey stone was once quarried in Asia Minor, then imported to France where it was cut and finished and exported all over the world as hones for cutthroat razors. For years this was the stone craftsmen sought. It is of a softish nature and wears away quickly in use. It was available once in vast quantities, and is often to be found at car boot sales and other second-hand sources.

Belgium stones are available and are used in the musical instrument-making trade for sharpening knives and other small tools. This does not mean that the stone is only obtainable in small sizes. I have a piece that is 12in long, 2¾in wide and 1in thick (305mm x 70mm x 25mm). The Belgium is a fine stone that cuts at a reasonable speed. It makes a good finishing tool; I use it for wide plane irons.

MAN-MADE STONES

There have been many different man-made stones over the years. (Japanese man-made stones are discussed in Chapter Thirteen and what is said here does not apply to them.) Some stones have been made by finely grinding natural stone into a powder and then reconstituting it into a block. Many different abrasives have been used. Today, new man-made stones are of sound quality and reasonably priced. Unless a second-hand stone is offered at a give-away price it is not worth contemplating.

The stones to be found in nearly every retail shop are those manufactured by Norton. There are two different varieties of abrasive used: silicon carbide stones, marketed as Crystolon, and aluminium oxide, marketed as India. Crystolon stones are lower priced than India stones, and inferior to them. Both stones can be obtained in coarse, medium and fine grades. They are also made double-sided, with fine grit on one side and coarse grit on the other. The India stones are impregnated with oil by the manufacturer.

Other lesser-known manufacturers make stones, all very similar to those made by Norton. Many are priced lower than a Norton. The stones made by Williams Wheels (*see* page 146) can be recommended. As you only get what you pay for, I would recommend the Norton or Williams Wheels stones where available. I find that all the man-made stones are on the coarse side. They cut very quickly, but I would recommend a fine grade of natural stone for finishing the sharpening process.

HOW BIG?

The normal size of a bench stone is 8in x 2in x 1in (203mm x 51mm x 25mm), though some very expensive stones have the thickness reduced to as little as ½in (13mm). Providing the stone is

Fig 4.1 Oilstone case.

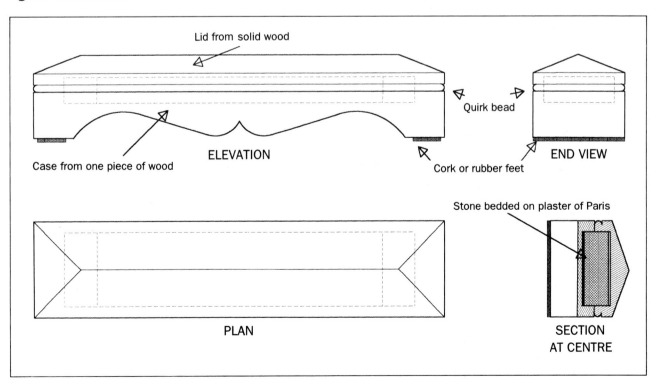

Lid from solid wood

Quirk bead

Case from one piece of wood

ELEVATION

END VIEW

Cork or rubber feet

Stone bedded on plaster of Paris

PLAN

SECTION
AT CENTRE

mounted in its case correctly this thinness is of no account. It might be supposed that the stone will wear out more quickly than a thicker stone. This may be so, but as the thin stones are always extremely hard they will last longer than the lifetime of a craftsman, even if used daily by a professional. Occasionally a stone that is somewhat larger than normal is to be found. A wide stone can be of particular use for wide plane irons. There are also many old razor hones still to be found second-hand. These are only an inch (25mm) wide, but are around 8in (200mm) long. While this narrowness means that less stone is presented to the tool being honed, a razor hone is still very useful.

PUT IT IN A BOX

The surface of the stone should be protected from workshop dust and other debris, and it is traditional to put the stone in a fitted case. This is normally made by letting the stone into a solid block of wood. A lid is made in a similar way (see Fig 4.1). There is a feature which I incorporate in my stone cases that you may find useful. At each end of the stone, a piece of hardwood is inserted. This is about an inch (25mm) wide and the same section as the stone. The grain in this block runs vertically. When honing, the blade or guide can overrun the stone by going out on to this block. This encourages one to use the whole of the stone's surface. Thus the stone is worn away more evenly, and truing has to be carried out less frequently. Some craftsmen make double or treble cases so that their stones are stored side by side. I do not recommend this for two reasons. The extra size of the case makes it more difficult to store than several single cases. Worse still, when backing off, that is to say when rubbing the back of the tools flat on one stone, the other stones are in the way.

Some stones are very fragile and are easily fractured. If these are mounted in a case properly, they will survive all the strains imposed by fair use. When making the case, the stone should be an easy fit in it. Once you are satisfied with the fit and are ready to install the stone, tip a small quantity of superfine plaster of Paris into the bottom of the case. The mixture needs to be quite thin, about the consistency of fresh cream. Gently press the stone down into the mixture until it is forced up around the sides of the stone. Put the case and stone to one side while the plaster sets hard. Once this process has been carried out, the stone should not be removed from the case. In fact, it will probably be impossible to do so. The stone bedded down on the plaster is firm and solid. This supports the stone throughout its entire length and makes honing more positive.

If you have a sharpening area (as suggested earlier) away from the bench, fix a lath across the stone area. If the end of the stone case is pushed against this, the stone will not move about when being used. Some craftsmen put panel pins into the underside of the case. These pins are cut off just proud of the surface and sharpened. The spike so formed grips the surface when the stone is in use. This might be fine for some people, but I find it spoils the surface of the bench if the stones are used on it. It is far better to fix a strip of cork or rubber at each end of the base, which is much kinder to the bench top.

STONE MAINTENANCE

No matter how carefully you try to use the whole surface of the stone, eventually it will wear hollow and need truing (known as dressing in the trade). This is done by rubbing the stone on a flat surface using an abrasive material. In most workshops the flat surface is an old piece of plate glass. The abrasive material is usually builders' sand lubricated with water. I have used an old fire clay kitchen sink with the sand and water covering the bottom, but not all stones require such drastic measures to dress them. The softer stones are easily flattened using a piece of coarse grade aluminium oxide paper taped to a flat surface. The harder the stone the longer the flattening process takes. This is a task that needs doing when one has the time; it is not something that can be rushed. In recent years I have used a coarse emery powder instead of sand. This imparts a much kinder surface to the newly flattened stone. The emery grinding paste, sold to grind the valves in car engines, can be used if emery powder is not available.

A newly dressed stone will be found to cut much faster than it did before dressing. This is because a new layer of sharp abrasive grains has been exposed. Some stones become clogged up with fine metal particles and/or congealed oil. This should not happen if the stone is used properly, but from time to time one comes across a stone in this state. A prime cause is that linseed oil has been used (wrongly) to lubricate the sharpening. The oil has oxidized and the stone no longer cuts freely. The old remedy, and one that I have never known to fail, is to boil the stone. This needs to be done carefully if the stone is not to be damaged. A can large enough for the stone to lie flat in the bottom is required. The stone is put in the can with a piece of rag underneath to support it. Cold water is added, sufficient to allow room for it to boil without boiling over. The can containing the stone and water is heated and left to boil for an hour or so. Heat softens the oil and the stone expands. The internal pressure of this expansion forces the now runny oil from the stone. Take the can from the heat and allow the whole lot to cool down before removing the stone. Do not remove the stone and try to cool it by placing it in cold water, as this is likely to crack the stone. If the stone is in a very poor condition it might require a second boiling, but I have never known this to be necessary.

LUBRICANTS

If it is liquid and freely available the chances are that a woodworker has at sometime put it on his sharpening stone. This cavalier attitude may have

Fig 4.2 Correct position of hands when honing a plane iron.

come about through experimentation, or perhaps there was nothing else readily available. In some workshops there are all sorts of concoctions, but I have yet to find anything better than the prescribed lubricant, be it water or oil. However, the title for both liquids covers a range of compositions. Some stone suppliers sell a special honing oil. This is expensive and has very little merit over other oils of the same consistency.

Certain stones that are lubricated with water, cut better if a small quantity of detergent is added to the water, but this is not always so. I know of several craftsmen who use a mixture on their oil stones. This comprises 90 per cent water, one per cent washing-up liquid and nine per cent soluble oil, the type engineers use as a cutting coolant. It is claimed by the users of this brew that it allows the stone to cut freely, clears the metal from the pores of the stone, and does not make the mess oil does. Old woodworking books often recommend neat's-foot oil thinned with about one-tenth its volume of paraffin. Neat's-foot oil is a light yellow oil, obtained from the feet and shin bones of cattle. Sweet oil is also recommended in some old books – I think this refers to olive oil. Oil can range from paraffin (kerosene) to thin grease. Water can be distilled or taken from a ditch. So just what should you use?

First we must ask what we require of a lubricant. The major task of this liquid is to keep the surface of the stone free from the small particles of steel removed from the tool being honed. This prevents the surface from becoming clogged up. (It is what engineers call a cutting compound.) Therefore we require a free-flowing, thin liquid, but not too thin or it will all run off the stone. Observe what is happening to the film of liquid on the surface of the stone during honing. Is it clearing the surface of metal particles? Is it staying on the surface of the stone with all the metal particles suspended in it? If so, it is doing its job; if not, modify it. If ordinary engine oil is thinned with a small quantity of paraffin the viscosity can be adjusted until you have the ideal mix.

Fig 4.3 Wire edge.

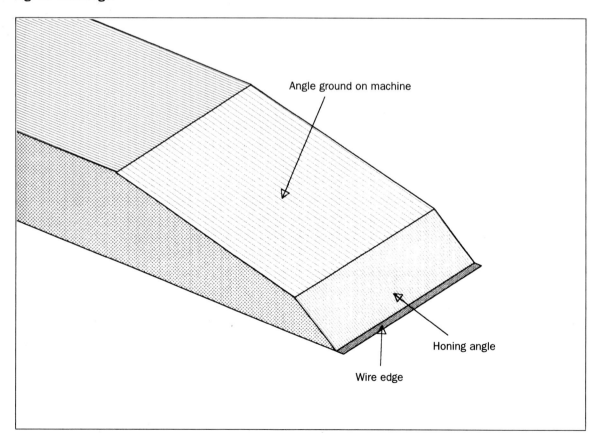

Angle ground on machine

Honing angle

Wire edge

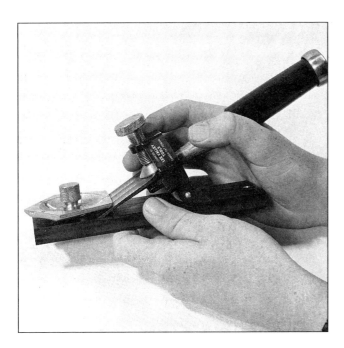

Fig 4.4 Setting the angle on a Lee Valley honing guide.

Fig 4.5 Side view of the Lee Valley honing guide.

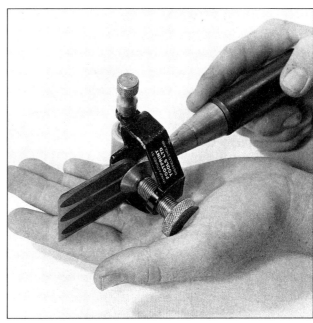

HONING ON THE STONE

The great difficulty the novice experiences when honing is keeping the bevel flat on the surface of the stone. That is, keeping the angle of the iron to the stone constant throughout the operation. As the tool is moved back and forth along the length of the stone there is a tendency to move the angle up and down. There is also a propensity to incline the tool at a steeper angle when it is close to the body than when it is at the end of the stroke. To counteract this, the hand should be raised slightly at the end of the stroke and the elbows should be kept close to the body so that all the movement comes from the shoulder and elbow. The wrist is locked and kept at one angle and the fingers of the right hand should lay down the length of the tool with their tips near the cutting edge. The left fingers should be placed across and on top of those of the right hand (*see* Fig 4.2).

When moving the tool back and forth along the length of the stone, try to use the whole surface of the stone. Some craftsmen advocate a figure of eight movement when honing narrow chisels. I do not advise this, because the stone will end up getting much more wear at the centre of the '8' than anywhere else. It is far better to use a straight back and forth movement and gradually work across the width of the stone. When you think that it has been honed sufficiently, inspect the edge for the candle (*see* pages 3–4). There should not be one. In its place a **wire edge** has probably formed. This will depend on the grade of the stone being used (*see* Fig 4.3). Remove this wire edge by drawing the blade through a piece of end grain deal. The honing then continues on progressively finer stones until the stone being used does not produce a wire edge. This final stone should produce flat, shiny surfaces that appear as if chrome-plated. When you have finished, wipe the soiled oil from the stones and replace the case lid.

HONING GUIDES

While a honing guide will keep the tool being honed at a set angle, its use is frowned upon by many craftsmen. The feeling is that anyone worth their salt ought to be able to hone properly without this tool. I am of a different opinion. Having experimented by using several different appliances, mainly while instructing apprentices, I have decided that a better edge can be obtained with the guide than without it. I have even gone to the length of designing and manufacturing a special guide for plane blades. If you wish to learn to sharpen without a guide it is a mistake to ever use one. As it was once explained to me, if you want to learn to drive a car you do not employ a chauffeur. One of the problems with many proprietary guides is the narrow roller, which wears the centre of the stone. I have tried out several guides and have found two which are commercially available that I can recommend with certain reservations.

The Lee Valley tool, made in this country by Footprint Tools Ltd, is of a very good design. This instrument has a unique feature that allows the angle to be adjusted 1° either side of an initial

Fig 4.6 Eclipse 36 honing guide before modification.

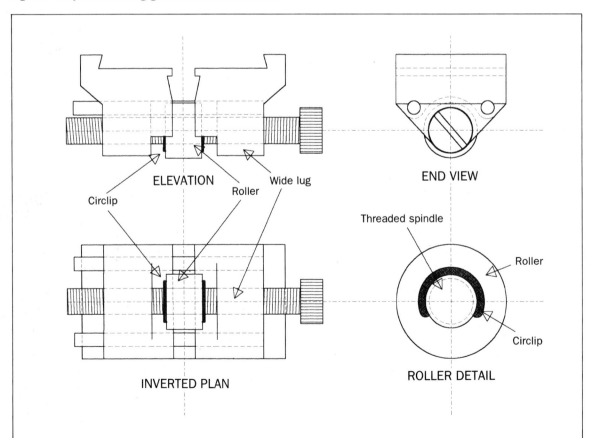

40

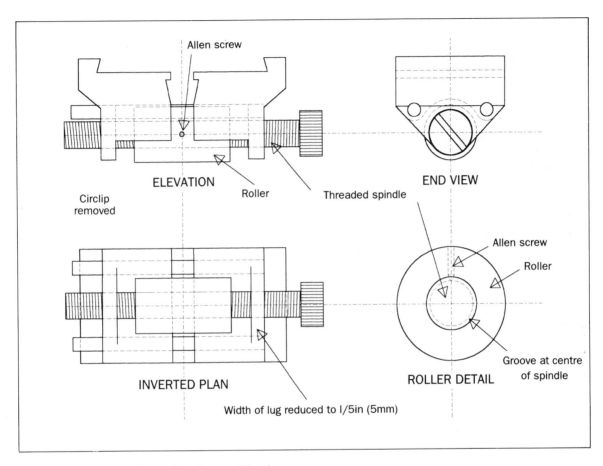

Allen screw

ELEVATION

Circlip
removed

Roller

Threaded spindle

END VIEW

Allen screw

Roller

INVERTED PLAN

ROLLER DETAIL

Groove at centre
of spindle

Width of lug reduced to 1/5in (5mm)

Fig 4.7 Eclipse 36 honing guide after modification.

**Fig 4.8 Eclipse 36 honing guide, showing my
modification to the roller.**

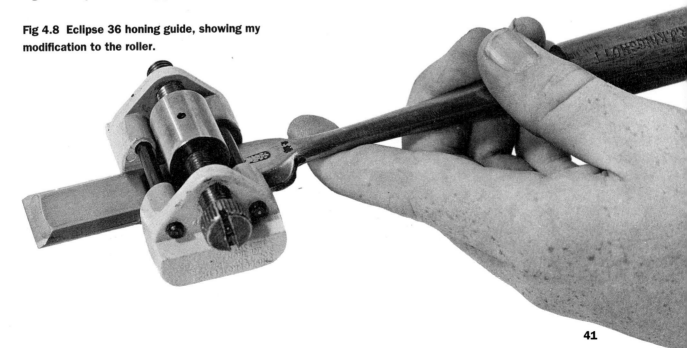

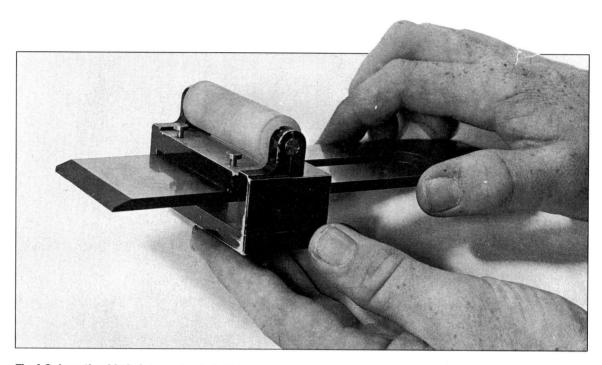

Fig 4.9 Inserting blade into my honing guide.

Fig 4.10 My honing guide in use.

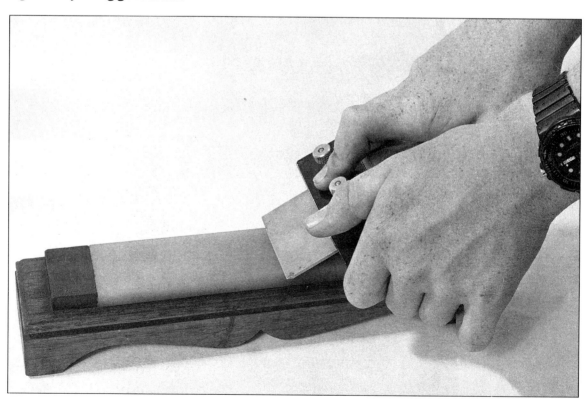

setting. This is ideal for applying a micro bevel. There is one major fault with this tool: the main frame is not stout enough. With continuous use the lower part is distorted until it impinges on the roller. This stops the roller from rotating and a flat is quickly formed. If the user is aware of this defect and the tool is not clamped too tight, the frame can be bent clear of the roller before any damage is caused. How long the tool would last with this periodic bending one can only guess at. The tool is expensive as guides go; perhaps there will ultimately be a modification. I have replaced the roller on mine with one made from PTFE (polytetrafluoroethene), which is much kinder to the stone (see Figs 4.4 and 4.5).

The Eclipse honing guide, which holds tools from ⅟₁₆in (1.5mm) to 2⅝in (67mm) wide, is a useful item that is suitable for most tools. I have had one for years and I use it for most of my chisels and plough irons. I have carried out a small modification to install a wider roller (see Figs 4.6, 4.7 and 4.8), which I think improves the tool a great deal. The wider roller makes the tool more stable on the stone. The extra width spreads the wear over a greater portion of the stone's width.

Figs 4.9 and 4.10 show the honing guide that I designed and made. By studying Fig 4.11 it is possible to make one, or have one made. There is no copyright on the design. You are free to use it should you wish.

Fig 4.11 My own design of honing guide.

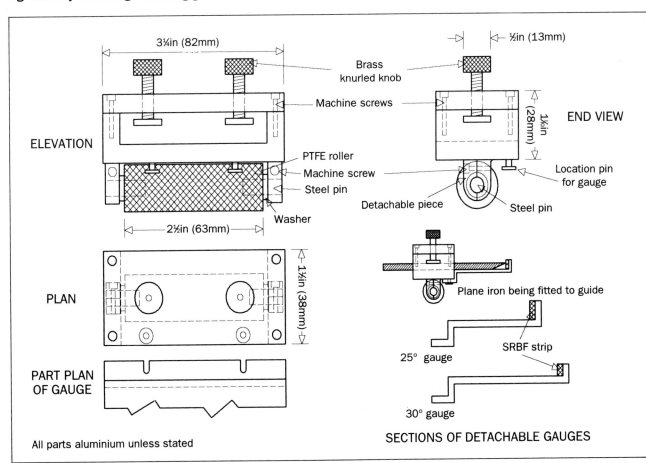

CHAPTER 5

DIAMOND SHARPENING SYSTEMS

I WOULD NOT BELIEVE IT

If ten years ago somebody had told me I could have a bench stone that would remain flat no matter how much I used it, I would not have believed them. Had they told me that the stone would cut faster than anything I already had and it could be a foot long and over 2½in wide (305mm x 65mm), I would have laughed at them. If they had gone even further and said that the grade of stone could be chosen from coarse to extra fine, and the grade guaranteed over the whole sharpening surface, I would have known they were pulling my leg. Today we woodworkers can have a stone that fits this description. Ours might be the oldest craft in the world, but there is no reason why we cannot benefit from modern technology.

One has to be very careful when reading the advertising blurb put out by manufacturers. Very few of the people who make or sell woodworking equipment are woodworkers. I find woodworking exhibitions most entertaining at times when I listen to the rubbish most of the exhibitors impart to the public, and watching the clumsy way some tools and machines are demonstrated is even more amusing. Such experiences have made me quite cynical, particularly where new and revolutionary things are claimed for the latest gizmo.

For some years I worked with a group of apprentices: young people with fresh, receptive minds. They were prepared to accept new ideas and methods that a staid old codger like me would

not look twice at. As we were employed by a large research and development establishment, things could often be obtained under the pretext that they were needed for evaluation. That is how the apprentices persuaded me to get a **diamond-surfaced sharpening plate**.

While this item abraded steel away very fast, it was not flat. There was a good ⅟₁₆in (1.5mm) hollow in the 8in (203mm) length of the thin plate that was bedded in plastic. The plate was rejected and we returned to using our trusted oilstones. At the next woodworking exhibition, there was a stand with the same sharpening plates being displayed. This delighted the apprentices as they could now tell the manufacturer what a lot of rubbish his product was. The gentleman on the stand listened politely to their complaint, then he picked up the item he was demonstrating and turned it upside down. Embedded in the plastic were two hollow square section metal bars running the full length of the plate. We were told that the manufacturer had soon realized that the original tool bent and twisted, and the metal bars had now been inserted to keep it permanently true. My name and address were taken and a few days later I received a new plate with the bars in it. The apprentices used this constantly, and so did I, right up until the establishment was recently closed.

I have related the above to show the reader who might be sceptical about what is claimed for this system that I too would not accept it at first. However, I now have four plates from two different manufacturers in my own workshop and I have no reservation in recommending them. Not only do I use the plates as I would a bench stone, but I also have several small diamond-faced tools which are used similarly to slip stones. However, I must say that these diamond tools have not completely replaced my stones. I still use Japanese water stones to put that last bit of magic on a cutting edge.

IT IS HARD

Diamond is the hardest material known. As it is a crystal it has sharp points and edges. It will cut any material: steel, ceramics, glass and, of particular benefit to woodworkers, tungsten carbide. Most stones need their surfaces rejuvenating from time to time, but because the cutting edge of the diamond remains sharp, the diamond sharpening surface does not require this. It seems to be maintenance free. Most of my diamond tools have been in continual use for several years. Apart from keeping them clean, there is nothing else that one need do.

There are two different surfaces available: one has a diamond-coated, perforated metal plate embedded in plastic; the other is a flat metal plate

Fig 5.1 Diamond sharpening systems.

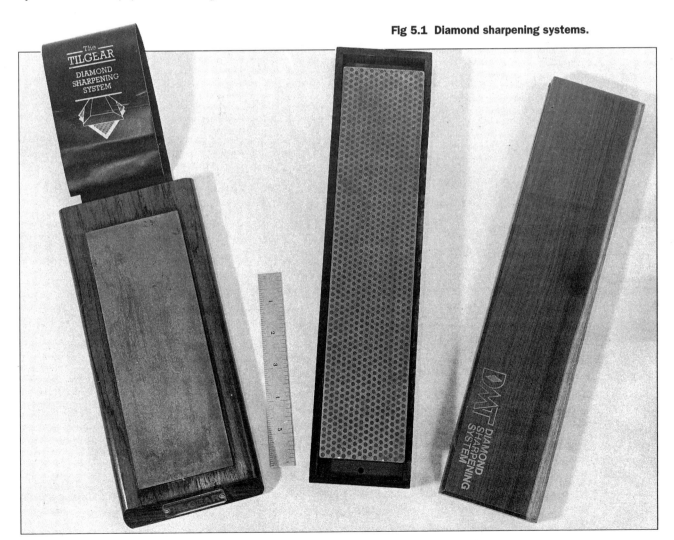

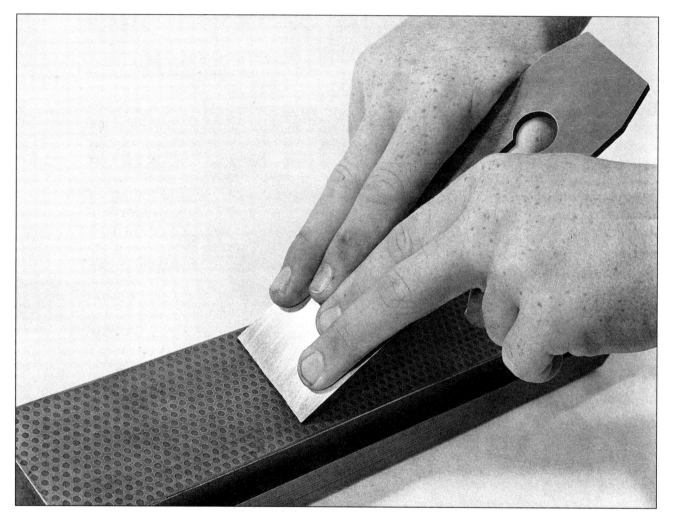

Fig 5.2 Sharpening on a diamond plate. Note the position of the fingers; this is not as I recommend, but is a position used by some craftsmen.

¼in (6mm) thick and covered with diamond on one face. I have both and can find little difference between them. The perforated plate has fewer diamonds due to the smaller surface area of metal. However, it cuts as quickly as the plate that has the whole surface covered with diamonds.

There is no agreed terminology when referring to these sharpening systems. They can hardly be referred to as a stone, even if the diamond is a stone. Is a rod of metal covered with diamonds a

file? Can the small diamond-coated plate be called a slip stone? I decided that as the diamond-coated tool used in place of the bench stone is a plate of metal coated with diamond, I'd call this a plate. Diamond Machining Technology Inc. (DMT), who manufacture some of these tools, have registered the name Diamond Whetstone as a trade mark. This confuses the issue even further. No doubt as the diamond sharpening system becomes established and accepted by the craft, appropriate names for all the new products will be agreed upon.

THE SYSTEMS

The sharpening system with the perforated plate embedded in plastic and reinforced by two metal bars is marketed in this country by Starkie & Starkie, (*see* page 146). It consists of a layer of monocrystalline diamonds, electrolytically deposited with nickel on to the surface of a steel plate. The plate is embedded in an injection-moulded polycarbonate base, reinforced with metal. This is accomplished with a 160 ton hydraulic press. The diamonds are securely held with about two-thirds of their size in the nickel. The flat metal plate is lapped before plating. The manufacturers claim a tolerance of 0.001in (0.025mm). The larger 'Superflat' plates are claimed to have an even better flatness tolerance of 0.0005in (0.013mm).

The sharpening system is fast, which is to say it abrades metal away very quickly. The DMT system has four grades of stone. These are colour coded which is very convenient in use. Green is extra fine, red is fine, blue is coarse and black is extra coarse. These tools are not cheap; at the time of writing a 12in (305mm) plate will set you back over £100 before VAT. As with all good ideas there are copies; remember there is nothing that someone will not make a little worse and sell a little cheaper.

In addition to the flat plate used in place of an oilstone, there are numerous other forms of diamond-coated tool made by DMT, and many of these have applications in the woodwork shop. There is a series of tools that can best be described as files. They have plastic handles made in two parts which fold around the tool when not in use. It is possible to restore the edge on TCT tools with these. Not only will this ensure that the edge is sharp, it can save a lot of downtime. Some craftsmen have to keep duplicate router cutters so that when one is away being ground, work will not be held up. Think of the saving in cutter cost, that is if you are brave enough to attempt sharpening router cutters.

The round and half-round tools are of particular use when sharpening in-cannel gouges. These gouges have always presented a problem when they need grinding. The diamonds quickly form a new bevel inside these tools. I find these diamond-coated tools easier to control than using a gouge cone on the grinder. There is a honing strip that will fit into a hacksaw frame and is ideal for sharpening TCT circular saws. It is possible to obtain a sheet of thin brass shim stock that has been plated with diamonds. This can be cut and bent, or stuck to shaped blocks of wood to make special shaped slips.

There is a use for diamond-coated tools that, as a cabinetmaker, I have a particular interest in. This is their ability to smooth the edge of glass. Cutting glass to fit in the shaped astragal bars of a bookcase or similar piece of furniture has always been difficult. Now I just nibble the edge with a pair of pliers until it is the right size and shape, then true it up with the diamond-coated tool. Not only does this save me time but also there are less bits of broken glass around the workshop.

The solid plate of steel with a diamond surface is made by Eze-lap Diamond Products in the USA and marketed in this country by Tilgear (*see* page 146). It is available in several sizes. The 8in x 3in (203mm x 76mm) plate that I use is mounted on a piece of hardwood and has a leather cover that is fastened by a velcro strip. This is one of my favourite tools. Tilgear also supply a 6in x 3in (152mm x 76mm) unmounted plate in a leather case which I keep especially for grinding the face of Japanese chisels and plane blades (*see* pages 135–7). Tilgear also markets a plastic-handled flat diamond file that has a cutting surface 2in long and ¾in wide (51mm x 19mm). The file, which can be obtained in three grades of abrasive, is useful for sharpening router cutters and suchlike.

LUBRICANT

DMT issue quite a good instruction sheet with their tools. This states categorically that water should be used on the surface when sharpening. They say that the islands of plastic that are formed by the holes in the perforated plate, work with the water to keep the sharpening surface clear of the particles of metal removed from the tool. I have found that using water, both on this system and with water stones, leaves a surface on the tool that very soon rusts. To avoid this I wipe the tool dry, making sure that all traces of water are removed, and then I wipe the tool with an oily rag; I use 'camellia oil'. The instructions that are included with the Eze-lap plate do not mention lubricant at all. I use a thin oil, which seems to work very well.

The cutting characteristics of all diamond-coated tools change after they have been in use for a while. At first they are extremely abrasive – almost vicious – they literally scratch the metal away. I am told that this is because there are many loose diamonds that are not properly embedded in the plating material. Once these are removed by the sharpening process the tool settles down and sharpening is more controlled.

While the diamond sharpening system is good, it in no way replaces the traditional tools in my workshop. Perhaps if I were just starting in the craft my feelings would be different, but I find that a good water-cooled grinder and a series of stones still give me the edge and the working characteristic I require. However, the diamond plate is so flat – and can be trusted to be flat – that it has become a highly important adjunct to my other sharpening tools. As for the diamond tools that are used to abrade tungsten carbide, they have no equal, and I would find it difficult to manage without them.

The main problem for the person wishing to try out the diamond system is likely to be the cost. With most sharpening stones one knows what to expect. The things have been around for a long while and there is plenty of unbiased information available on them. Added to this they are relatively inexpensive. One is not normally as ready to spend money on a fairly new and unknown product, particularly if it costs a lot more. I would say as an experienced woodworker that an 8in (203mm) plate would be very useful in any woodworking shop. It will also save quite a large amount of the time spent on sharpening bench tools. However, when it comes to machine tool cutters, the knowledge and experience of the craftsmen should be the deciding factors (*see* pages 116–123).

CHAPTER 6

STROPS AND STROPPING

A PROGRESSION

Having read thus far, you will have realized that obtaining a sharp edge is nothing more than abrading metal away using progressively finer abrasives. Stropping is another step in this chain. It consists of rubbing the tool on a piece of leather impregnated with a very fine abrasive. The abrasive is sold as stropping compound and is obtainable in two grades of grit. As the strop is made from leather it is not possible to rub the tool back and forth on its surface as one would when sharpening on a stone; the tool would cut into the surface of the leather on the push stroke. Instead, the tool is drawn back along the strop and then lifted clear of it on the return stroke. Unless a superfine sharp edge is required a strop is not used.

It should be remembered that the very sharp edge of a tool only lasts a few minutes. In Chapter One I explained that a sharp edge is where two flat planes meet, and that the object is to get the metal where they meet as thin as possible. How long will this extremely thin piece of steel at the very edge be able to stand up to being constantly pushed into a piece of wood? The answer depends on many variables, such as the quality of the steel, the hardness of the wood, the force used to propel the tool, the angle of the cut in relation to the grain, etc. This is why the strop is very useful. It can be kept close to the job in hand and as soon as it is felt that the keen edge of the tool is going, it can be pulled across the strop a few times, and the superfine sharp edge is restored. This is particularly useful when carving or paring with a gouge or chisel. To strop a plane iron

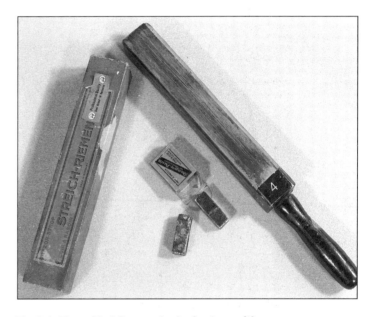

Fig 6.1 Four-sided German barber's strop with packet and sticks of stropping paste.

it has to be removed from the plane. So the strop is not often resorted to when planing.

WHAT IS A STROP?

A strop is nothing more than a piece of leather. The shape and size of the strop can vary from something that resembles a leather boot lace, for use on the inside of very small gouges, to a chunky 12in x 4in (305mm x 102mm) glued flat on a piece of ply. Most strops are double-sided, and have different abrasives applied to each side. There is also a commercially made strop available from some tool shops. This tool comes from

Fig 6.2 A selection of my slip stones. The two marked with an 'A' are leather-covered wooden blocks.

Germany and is made for the barbering trade, but it is very useful in the woodwork shop. It is four-sided, and has three faces covered with leather. The fourth side has a thin piece of Turkey stone on it (see Fig 6.1). Barbers obtain these strops from a wholesaler so, if you have trouble finding one, it might be a good time, when you are next having your hair cut, to see if the barber will order one for you. The most used strop in my workshop is a bit of ⅜in thick mahogany, 12in long and 3in wide (10mm x 305mm x 76mm), with leather on both sides, and a hole drilled at one end so that it can be hung up close to where it is needed. I use several shaped strops for sharpening in-cannel

gouges. These are nothing more than a shaped block of wood with leather glued around it (see Fig 6.2). The leather used for making strops is thick hide that has been dressed smooth on one side. There is some discussion as to which side of the leather to use for stropping on. I prefer the smooth side, but I know craftsmen who use the other side; this soon becomes smooth with use, so what is the difference?

STROPPING PASTE

The abrasive used on strops comes in the form of a stick, which resembles a wax crayon. Again, the main manufacturer is in Germany. There are two square sticks packed in a small cardboard box, about half the size of a match box; one is red, the other black (*see* Fig 6.1), yet the abrasive quality of both is so mild as not to be discernible when rubbed between thumb and forefinger. While the size of the stick is small, they last for a very long time, and the strop once dressed only needs occasional treatment with the dressing. Dressing is applied to the surface of the strop just like scribbling with a crayon. The first time a new piece of leather is dressed it requires quite a lot of the paste. After that only a small amount is needed. The tool being stropped spreads the paste and drives it into the leather. A new strop does not work as well as an old one.

HOW TO STROP

Stropping is probably the easiest of all the sharpening techniques to master. It is best to practise with a short chisel about ¾in (19mm) wide. The tool must be sharp to start with; the strop only adds that last little something that makes the tool super sharp. Stropping is similar to putting a micro bevel on a tool (*see* page 15). This is because the tool's bevel is not applied flat on the strop. To start stropping, lay the bevel flat on the strop and then raise the tool a couple of degrees so that only the edge is in contact with the leather. Now, while applying moderate pressure, draw the tool towards you. For flat tools such as chisels and plane irons the strop can be placed flat on the bench, but some craftsmen (including myself) prefer to put one end on the bench and support the other in their left hand. Once the bevelled side of the tool has been drawn over the strop about six times, it is turned over and the flat side is treated similarly. When it is being stropped, the flat side of the tool is not put flat on the strop; the end of the tool is raised very slightly so that only the edge is in contact with the leather.

SPECIAL STROPS

Shaped tools need special treatment. The basic theory is the same but a shaped strop is used to get inside the shape. The tool is usually held firm against the bench while the strop is moved. A new strop will leave some of the stropping paste on the surface of the tool and this has to be wiped off or it will transfer to the work in hand. Once the strop has been used for a few days this nuisance does not recur, except when the strop is occasionally re-dressed. Because the stropping paste on the surface of the strop is waxy it is inclined to pick up dust and grit, which of course will spoil any edge that is dragged over it. To prevent this happening, most strops have a case in which they are kept. This is a very simple affair (*see* Fig 6.3).

Fig 6.3 Two-sided strop with case.

CHAPTER 7

Machine Sharpening

In General

The human animal is inherently lazy. He will spend countless hours trying to find an easier way of doing a task. This is particularly so when the task is one that he finds irksome. Let us face it, nobody enjoys sharpening their tools. It is only a means to an end; using the tools is what we all want to do. It is easy to keep putting off the moment when we have to stop productive work and turn our attention to sharpening. In view of this it is not surprising to find that some entrepreneur has tried to mechanize the whole process of sharpening.

There are a number of machines made that, it is claimed, can sharpen a tool. The problem is – as I discussed in Chapter One – how do you define sharp? Just what sort of working edge is required? I have never had much faith in the ability of a machine to give me the edge I want but, to enable me to write this chapter with some authority, I obtained several machines and appliances. These I have used to sharpen a range of tools. The edge has been examined with a magnifying lens and then the tool has been put to work in the way it is normally used. It must be emphasized that the comments on the results are my own. One's tools are very personal, and the normal sharpening techniques are modified to suit one's own particular way of working.

I have tried to obtain a cross section of all the machines available which a woodworker might find useful. While the manufacturer and type of sharpening machine are mentioned in most examples, this is only done so that the exact machine may be identified. There are, in most cases,

similar machines made by other manufacturers and, added to this, the design of machines is constantly changing. Therefore it is the method of obtaining a sharp edge that I have concerned myself with; not the machine. I have, however, commented on a machine's construction when I have felt it may be helpful. The primary purpose of this chapter is to describe the various ways different manufacturers have approached the problem of machine sharpening. All sharpening techniques require practice to achieve the best results. I have spent at least four hours with each system described here. Some experimentation has been carried out, e.g. pressure applied to the tool while honing. Other variables such as the amount of abrasive dressing applied have been investigated. It was found that very light pressure on the tool gave the best results. The amount of dressing did make a difference; when the surface had not been recently dressed, it gave the sharpest results. Some of the machines have been used in my own workshop on a daily basis. Others after a fair trial period have been rejected.

I suppose that the grinder could be classified as a sharpening machine, but what we are looking at here are machines that hone the tool. No use of the oilstone is made; the whole sharpening process is done by the machine. Most of these machines rely on a fine abrasive put on to a flexible base; the simplest is a straightforward fabric buffing wheel dressed with a fine abrasive compound. Most of our tools have bevels that are flat; these are obtained, and maintained, by rubbing the tool on a flat abrasive surface. The question is, can the flat be maintained if the tool is held against the edge of a wheel, particularly if the surface of the wheel is flexible, which is the case when a fabric wheel is used? When pushing the tool against the wheel, the flexible surface presses hardest against the ends of the bevel, which causes a rounded edge. Perhaps a round at the top of the bevel where it meets the unground part of the blade will not affect the tool's working properties, but the cutting edge, which is normally formed by two flat surfaces, is a very different story. The whole working characteristics of the tool will change if the edge is formed by rounded surfaces. The control of the cut when using chisels and all of the carving and turning tools is dependent on the cutting edge being formed by flat surfaces. If the edge is rounded, the handle of the tool will have to be raised to bring the cutting edge into contact with the work (see Fig 7.1), and control over the cut is then lost.

DIFFERENT TYPES OF MACHINE

Some machines get over the round edge problem by using an endless leather belt. The belt runs between two rollers, one of which is powered. The other roller is adjustable so that the belt's track can be altered. Between the two rollers the belt is supported by an adjustable platform, and the sharpening is carried out where the belt runs over the platform. In fact this could be described as powered stropping (see Chapter Six). The leather belt can be replaced by an aluminium oxide-coated fabric belt, making the machine similar to a linisher as used by engineers. Belts of various grade of grit size are obtainable. The problem with this type of machine is one of heat; care must be taken when holding the tool against the belt not to allow it to become overheated. These machines also have a dry grinding wheel. An acceptable edge was obtained using this method on firmer chisels and turning tools. Carving tools needed some handwork after machine sharpening to put the final touch to the cutting edge. I did not try to sharpen plane irons on the machine as the honing area was of insufficient size. The machine I used was an Elu MWA 149. I was impressed by the rigid construction and the jig for holding plane irons and chisels while grinding.

The Tormek machine, though principally a wet grinder, has a wheel which the manufacturers claim can be used for honing. This is accomplished by dressing the wheel with a flat stone, called a **grading stone** by the manufacturers. One side of the grading stone is very coarse and the other side is fine. When pressing the fine side of the grading stone against the grindstone, the surface of the grindstone is smoothed. It can then, they say, be used for honing. When the coarse side of the grading stone is pushed against the grinding wheel it reverts to a coarse-grade stone. This process can be repeated an unlimited number of times. The machine is also fitted with a wheel which the manufacturers call a **de-burring wheel**, used on the flat side of a blade to remove the burr. The machine has numerous attachments for sharpening a variety of tools, ranging from planer blades to scissors. The question is, does the system work? I found that as a wet grinder it was fine, but I was unable to get what I would class as a really sharp edge on either chisels or plane irons. However, it did reduce the time spent honing the tools by hand. The de-burring wheel has a flexible surface which is joined at one place in its circumference. On both wheels that I used the material either side of the join did not align properly, causing a pronounced jump in the tool every time it passed over the join.

Another method of machine sharpening is to use a fine abrasive wheel. These are usually bonded in rubber and must therefore always revolve away from the edge of the tool being sharpened. The Scangrind 200S supplied by Record uses this method. A wet grinding wheel of reasonable size runs in a water trough. and a rubber grinding composition wheel of smaller diameter can be adjusted to run against its

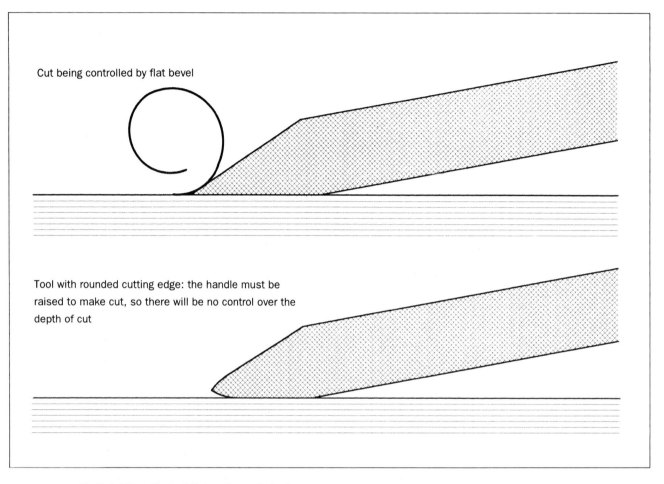

Cut being controlled by flat bevel

Tool with rounded cutting edge: the handle must be raised to make cut, so there will be no control over the depth of cut

Fig 7.1 The effect of flat and rounded edges on cutting characteristics. (This applies to any chisel, gouge, carving or turning tool.)

perimeter. This rubber wheel is used to hone the previously ground tool. There is some difficulty in holding the tool at the correct honing angle but, once this is mastered, the process is very straightforward. I obtained a good sharp edge on some carving chisels and wide Japanese chisels. The wet grinding part of this machine is the best water-cooled system I have used. The only fault I could find was that the tool rest moved due to the flexible nature of the plastic machine casing to which it is fixed. It is only possible to hone flat-bladed tools on this machine.

One machine that I tested came with no instructions (I think it is a D-6751 Eulenbis); it is manufactured in Germany by Kurt Kotch. At first sight it looks remarkably simple; it consists of an electric motor with an extended shaft on which are mounted two felt wheels. The tool rest consists of a block of wood, but I found it much easier to ignore the tool rest and hone the tools freehand. The wheels are made from two different grades of felt, one much coarser than the other. A block of abrasive dressing is supplied with the machine. The machine is aimed at the woodcarver, and for chisels and shallow gouges it gives an acceptable edge. However the edge obtained tends to be rounded and there would be a need to resort to the grinder occasionally to correct this fault. As with all woodworking techniques, some practice is needed before the best results are obtained. I found that a very light touch was required.

Ashley Iles supply a machine sharpening kit for woodcarvers. This consists of a rubber abrasive wheel, a stitched dolly mop and a block of abrasive dressing, accompanied by very explicit instructions for mounting and using this equipment on a bench grinder. The rubber wheel and dolly are attached in the same way as a normal grinding wheel (*see* Chapter Three). Unlike a normal grinding machine they need to revolve away from the operator, so the machine has to be turned around and the guards remounted. This did not take as long as I had anticipated. The rubber wheel needs truing after it has been mounted, which can be done with a stick of aluminium oxide. Following the instructions given, I obtained a good working edge on a variety of carving tools. The dolly mop, I discovered, is the only machine sharpening item that really gets inside a carving gouge (I only tried the system on shallow gouges). The green wax abrasive dressing worked very well in combination with the calico cloth of the dolly. Veiners and vee tools I would not chance on any method of machine sharpening.

Tilgear supply a leather disc with an arbour for mounting it in the chuck of a drill or lathe. There was also a very fine white abrasive dressing with it which could best be described as a mechanical honing system. I mounted it in a Jacobs chuck in the lathe and found that it worked best at the lowest speed of my lathe (around 300rpm). The leather disc, which is 3½in (89mm) diameter and

¾in (19mm) thick, worked best after it had been in use for a while; indeed, the surface seems to improve the more it is used. After several tools have been honed on the new disc it requires dressing with the abrasive. I found that by rubbing the surface of the leather after dressing with a small steel bar, the cut of the abrasive was reduced, which improved the edge of the tool being sharpened.

CONCLUSIONS

Having tried all these different methods of machine sharpening, I can say without hesitation that I can get a better edge by hand sharpening. Whether with more practice and experience of using machines this would change, I have my doubts. All the systems used tend to produce a rounded edge, particularly after the tool has been sharpened by this method several times. However, for certain uses there is a saving in time, which can make the method attractive. If you are buying a grinder and it has one of these systems as an addition then, without any argument, this is an advantage. I found the construction of some machines used to be on the flimsy side. Good solid castings, I know, cost more to produce than plastic mouldings or bent sheet metal, but there is nothing worse than parts of a machine that are supposed to position a tool accurately, flexing in use.

So, after trying out these machines, will I be using any of them on a regular basis? The Record Scangrind 200S and the machine by Kurt Kotch will have a permanent place in my workshop. The latter machine has been adapted to take the wheels from Ashley Iles as well as one of the original felt wheels.

CHAPTER 8

SHARPENING PLANE IRONS

MANY DIFFERENT TYPES

Plane irons come in all shapes and sizes and the same sharpening technique is not suitable for them all. A wide trying plane iron, for example, needs a totally different technique to a moulding

Fig 8.1 Sharpening angle stamped on Record back iron.

plane iron. Even blades designed to work on a flat surface vary; thick irons from wooden and English pattern planes and thin irons from Bailey pattern planes, can present us with different sharpening requirements. The irons in some older planes are laminated; that is to say, the iron is composed of two laminations of steel. The lamination on the face of the iron that does the cutting is made from high-carbon steel, and the rest of the iron is made from low-carbon steel. The latter, being much softer than the high-carbon, is easily ground away. The thin iron from the Bailey pattern plane is made entirely from high-carbon steel. Being much thinner than the conventional iron, the sharpening bevel is much narrower. It was originally suggested by the manufacturer that this is to save the chore of grinding the iron; however, I have a suspicion that it has nothing to do with grinding, but is rather a way of saving 50 per cent of the steel in the iron. Added to this, laminating is an expensive and time-consuming process. At one time Record stamped two lines into the back iron showing the sharpening angle (*see* Fig 8.1). In spite of what manufacturers say, most craftsmen still grind these thin blades.

A CONTENTIOUS SUBJECT

I have already said on page 5 that no two craftsmen seem to be able to agree entirely about sharpening techniques. There is particular disagreement about the angle at which plane irons are sharpened. Some craftsmen think the finer the bevel angle at which the iron is sharpened, the sharper the edge. Thus the plane cuts sweetly and

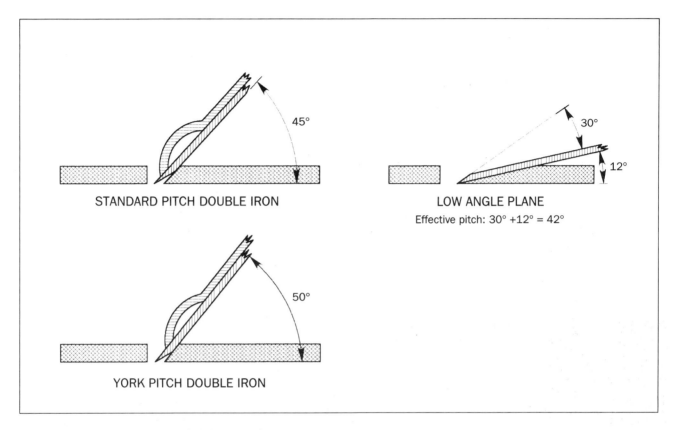

STANDARD PITCH DOUBLE IRON

45°

LOW ANGLE PLANE

Effective pitch: 30° +12° = 42°

30°

12°

YORK PITCH DOUBLE IRON

50°

Fig 8.2 How the pitch and bevel of the plane iron affects the cutting action. Here we see the effective pitch angle of the iron with bevel down and bevel up.

takes less energy to use. Others say that the bevel is behind the cutting edge and the only angle that affects the cutting action is the pitch of the iron (*see* Fig 8.2). If the latter is the case then the bevel is only there to give clearance to the cutting edge.

Whichever of these two views is correct, two angles have been established in the trade as the norm for most woodcutting tools. These are a grinding angle of 25° and a honing angle of 30°. Experimenting 5° either side of these angles is quite instructive. When you find the angle that suits your method of working best, stick to it despite what anyone else says or does.

MORE THAN JUST SHARPENING

A plane requires more than just a sharp iron to ensure that it works properly. There are many features in the plane's make-up that need inspecting, and perhaps modifying. Even a brand new plane straight out of its box will require this. Most manufacturers enclose a leaflet with the plane explaining some items that need fettling, but none of them goes in to all the points that need looking at. For our purpose I will take a new Stanley No.6 from its box and go through the preparation required to put it into use. This

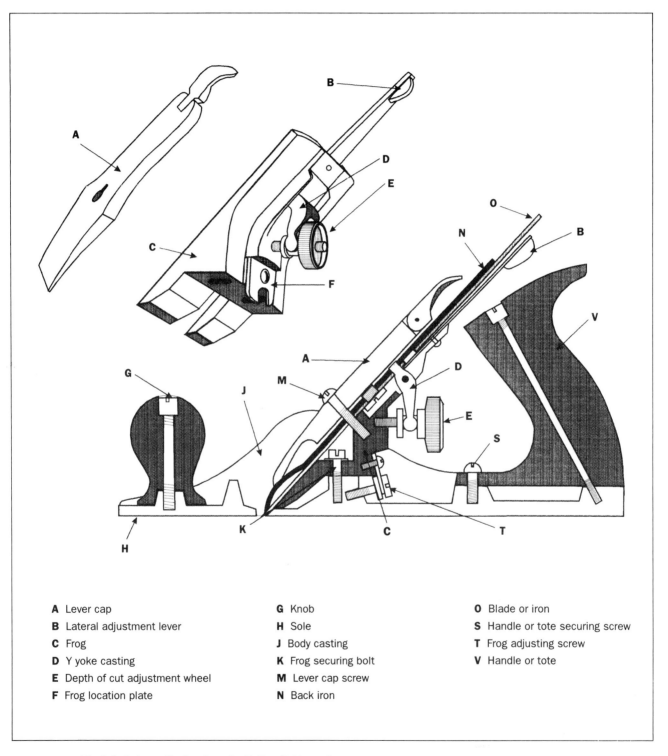

Fig 8.3 Schematic drawing of a Bailey Pattern plane.

A Lever cap	**G** Knob	**O** Blade or iron
B Lateral adjustment lever	**H** Sole	**S** Handle or tote securing screw
C Frog	**J** Body casting	**T** Frog adjusting screw
D Y yoke casting	**K** Frog securing bolt	**V** Handle or tote
E Depth of cut adjustment wheel	**M** Lever cap screw	
F Frog location plate	**N** Back iron	

preparation may not be considered as part of the sharpening technique, however, I have met so many frustrated novices who, having spent hours sharpening, cannot get the plane to cut properly, and it is clear that, if the benefits of a sharp iron are to be enjoyed, the initial setting up of the plane must be understood.

Fig 8.3 shows the names for the parts of the plane. As the Stanley No.6 is used to true the surface of wood it is important that the sole is flat. Even on planes not used for truing, the sole must touch the wood immediately in front of the mouth. The reason for this is explained in Fig 8.4. Remove the iron from the plane. With a good metal straightedge, check the sole for straightness along the fore and aft centre line, then position the straightedge diagonally (*see* Fig 8.5). This will show up any discrepancy in the sole. The chances are

Fig 8.4 Section through sole of plane showing different cutting actions of single and double ironed planes. The back iron and sole on the double ironed plane prevents tear out.

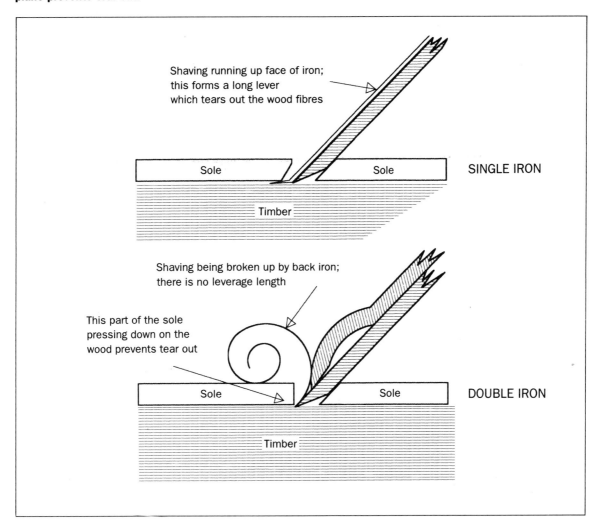

59

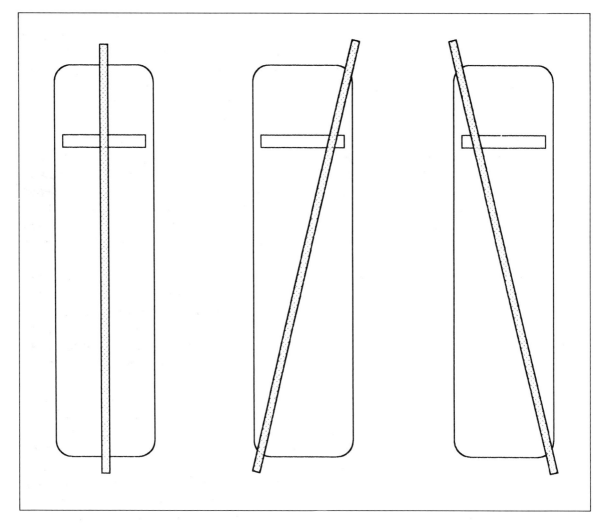

Fig 8.5 The three positions for the straightedge on the sole of a plane when testing for straightness.

that the sole will be hollow. This is caused by the casting moving after it has been ground. If the plane is to work properly any out of true in the sole must be corrected.

There is another point about the sole of new planes that needs looking at. Is the ground surface reasonably smooth? Are there marks from coarse grinding? The sole rubs on the surface of the wood, and if it is rough, it causes a good deal of friction. I once complained to a manufacturer about the standard of grinding on the soles of his planes.

The reply was: 'The coarse ground surface will hold paraffin wax, making the plane run sweetly'. Very few craftsmen put paraffin wax on the sole of their plane; the informed use an oil wick.

If the sole is at all out of true it can be put right using one of two methods. The first is to take it to an engineering firm and have it surface ground. The other is to rub the sole on aluminium oxide paper that has been stuck down on a perfectly flat surface. A piece of plate glass provides a very good surface for this purpose. The grade of paper depends on the amount of metal that has to be removed, but the final rubbing should be on paper no coarser than 150 grit. Now turn your attention to the iron. What is the face like? (The face is the flat side without a bevel.) Is it truly flat? Does it have grinding marks on it? First it must be rubbed flat on the surface of a stone to correct either of these discrepancies. Never try to true the iron by grinding it flat on the side of the grinding wheel. This may seem an attractive method of truing an iron, but be warned, it usually makes matters worse.

Once the face of the iron has been made true, and there are no signs of grinding near the cutting edge, fit the back iron. Hold the assembly up to the light. Can you see any daylight between the two where they bed together? The contact along the very tip of the back iron must be perfect, otherwise shavings will jam in the gap, and the mouth of the plane will become choked up. The best method of making adjustments to the back iron is to rub the surface that beds on to the tip of the blade flat on an oilstone.

The frog should now receive your attention. It can be removed by undoing the two screws securing it to the body. There are ground surfaces on the frog and the body of the plane to ensure the frog beds down solidly. Put a little engineers' blue on the ground surfaces of the frog and refit it to the bed. When the frog is again removed, there should be an even coating over the ground surfaces in the plane's body. A poor fit here causes the plane to chatter. Any discrepancy should be put right with a fine file. The frog is finally returned to the bed and adjusted so that the mouth is of a suitable size. Take particular care to ensure that the frog is set so that the blade is parallel with the front of the mouth. Now all that remains is to sharpen the iron.

THE SHAPE OF THE EDGE

Following the instructions on grinding and honing given in Chapters Three and Four, a suitable sharp edge can be put on the iron. However, the shape across the cutting edge will vary, depending on what the plane is to be used for (*see* Fig 8.6). Bear in mind, we are still talking about bench planes for use on flat surfaces.

The jack plane is the first to be used on a sawn surface that is to be trued. It is therefore required to make a heavy cut. The iron of this plane is rounded from side to side, cutting thicker at its centre than at the edges. A surface straight from the jack plane has slight ripples across it caused by the shape of the iron. Next is the trying plane. This is used to make the surface absolutely true. It would not do to have the rippled effect left by the jack plane. Therefore, the iron is sharpened straight across, with just the corners removed. If the corners were left square they would leave a step in the surface being planed. The jointer is sharpened absolutely square, as is the rebate plane. When it comes to the smoothing plane, used to finish the surface, special attention is required. The iron is sharpened square, then a very gentle round is applied to the corners. When the planed surface is examined, it should be impossible to find where the edge of the shaving was removed from. The tips of the fingers are far more sensitive than the eyes when inspecting a surface. Pass them lightly across the grain of the finished surface. Any discrepancy will be felt, and the shape of the plane iron can be modified.

It goes without saying that all planes used for working grooves and rebates must be sharpened with a straight cutting edge. This also applies to mitre and shoulder planes. Moulding cutters for combination planes are similar to moulding plane irons. Both are sharpened using slip stones. It might be thought when first examining the cutter or iron that many slip stones would be required to fit all these different shapes. Most craftsmen get by with four or five slips. The slip stone has a different contour along both edges, so four stones provide eight differently shaped edges. Cutters with a rounded (convex) end can be sharpened on a flat stone. Hollow cutters (concave) require a slip. The slip stone used should have a smaller radius than the iron it is being used upon.

A different technique is used with a slip stone than when sharpening on a bench stone: the tool is held still and the slip stone is moved to hone the edge. To keep the iron still it is best held against the side of the bench, with the cutting edge uppermost. Working in this way enables the edge being honed to be seen, and makes it easier to keep the sharpening bevel flat. For moulding cutters with more complicated shapes it may be necessary to use a knife-edge slip stone. While this tool has one edge that is extremely narrow, the other edge is a square about ¼in (6mm) wide. When working on cutters for combination planes, the profile of the edge may be changed slightly without affecting the working of the plane. The

wooden moulding plane iron is a different story; any deviation from the correct shape will preclude the tool from working properly. Some sloppy workers have been known to alter the profile of the plane's sole to fit a newly sharpened iron. Never resort to this; take care while sharpening and keep trying the iron in the body of the plane and sighting along the sole to see how things are going.

Planes used for trenching (a groove across the grain) have spur cutters fitted to sever the grain. These spurs need sharpening; they are bevelled on the inside and must be flat on the outside. Spur cutters wear out long before the plane blade, and it is possible to buy new ones (they are usually supplied in pairs). Some rebate planes have spur cutters fitted. These are only

Fig 8.6 The shape of the cutting edge of plane blades.

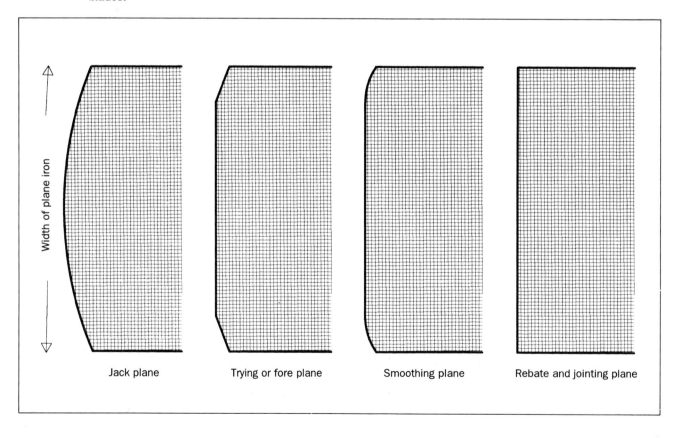

Width of plane iron

Jack plane   Trying or fore plane   Smoothing plane   Rebate and jointing plane

needed for cross grain working, and for normal long grain work they can be removed. Don't lose them; they are very small and easily misplaced.

SPECIAL PLANES

There are several planes with special uses that need careful attention when sharpening. The metal variety of side rebate has the blade sharpened at an angle across its width. This angle is important and must be maintained as it is very easy to sharpen a little more on one side than the other. It is good practice to try the blade in the plane several times while honing, to ensure the correct angle is being maintained. The skew mouth rebate planes have a similar iron and need

Fig 8.7 Honing the blade from a hand router on the edge of an oilstone.

the same careful treatment.

The toothing plane with its special iron is sharpened on the bevel side only. On no account should the face with the little grooves in it be rubbed on the stone. The edge that we are seeking on this tool is a row of evenly spaced sharp points. This plane is usually sharpened at a much steeper angle than bench planes. The iron is pitched almost vertical in the body, and the plane works with an action similar to scraping. The steep angle of the sharpening bevel adds support to the teeth. Toothing planes sharpened at a low angle tend to lose teeth, particularly on wood of a hard character.

Hand routers have blades that are necked. That is to say, the small cutting part of the blade is set at 90° to the rest of the blade. Because of this shape, sharpening is very difficult. The usual practice adopted by most craftsmen is to use the side of a bench stone (*see* Fig 8.7).

Once the general theory and practical methods of sharpening are mastered, any unusual plane that comes to hand can be dealt with. I would emphasize however that, as often as not, a plane is not working as it should because it needs some adjustment other than sharpening. The cause of most problems can be found by careful observation while using the plane.

Lastly there is the question of how far back from the cutting edge the end of the back iron should be set. This depends on the thickness of shaving and the type of timber being worked. Remember, the purpose of the back iron is to break the shaving up as it is removed from the surface of the work piece. A fine thin shaving needs a setting of ⅟₆₄in (0.4mm) or less, whereas a thick shaving from a jack plane would have a setting as big as ⅛in (3mm). If you find the surface grain is tearing out, reduce the setting. Quite often, Bailey pattern planes require some work inside the body along the front of the mouth. This is because when the mouth is set fine and the back iron is close to the cutting edge, there is no room for the shaving to escape.

Care of The Plane

Having spent time and effort getting the plane iron into tiptop condition, it would be a shame if we spoilt it through carelessness. How often one sees a plane standing flat on the bench with the iron in contact with the bench top. Some workers lay the plane flat on its side so that the cutting edge is not touching the surface of the bench. However, it is now exposed, and can be knocked by any other tool on the bench. The informed craftsman always has a strip of thin wood on his bench, placed level with the bench stop. If the front of the plane is placed on this strip, the plane is in the most convenient position for immediate use, and the cutting edge is in no danger.

When wooden planes, such as moulding planes, etc., are not to be used for several weeks they are stored with their wedges in a loose condition. The continuous stress put on the plane's body can, over a period of time, distort it. Second-hand planes are often found to have hollow soles caused by being stored with a tight wedge.

All metal tools benefit from the occasional wipe over with an oily rag. Only a very fine, barely detectable film of oil is required. Wooden planes develop that lovely patina from being occasionally wiped over with linseed oil. Do not overdo the linseed, however, otherwise it congeals on the surface and looks a horrid mess. Some workers suffer from damp hands, and the palms and fingers of their hands leave a deposit on metal tools that soon rusts. The only remedy is to give the metal surfaces a wipe at the end of every working period.

Plane irons are prone to corrode where the back iron tip beds against them. This is caused by dampness from the wood being deposited there. There is one bad habit that some woodworkers develop: when they wish to clear dust from the plane's mouth, they fill their lungs and blow it away with the expelled breath. This deposits all the moisture from their warm breath on the cold metal surface. A much better idea is to keep an old 1in (25mm) paint brush handy to brush away the dust.

CHAPTER 9

SHARPENING CHISELS AND GOUGES

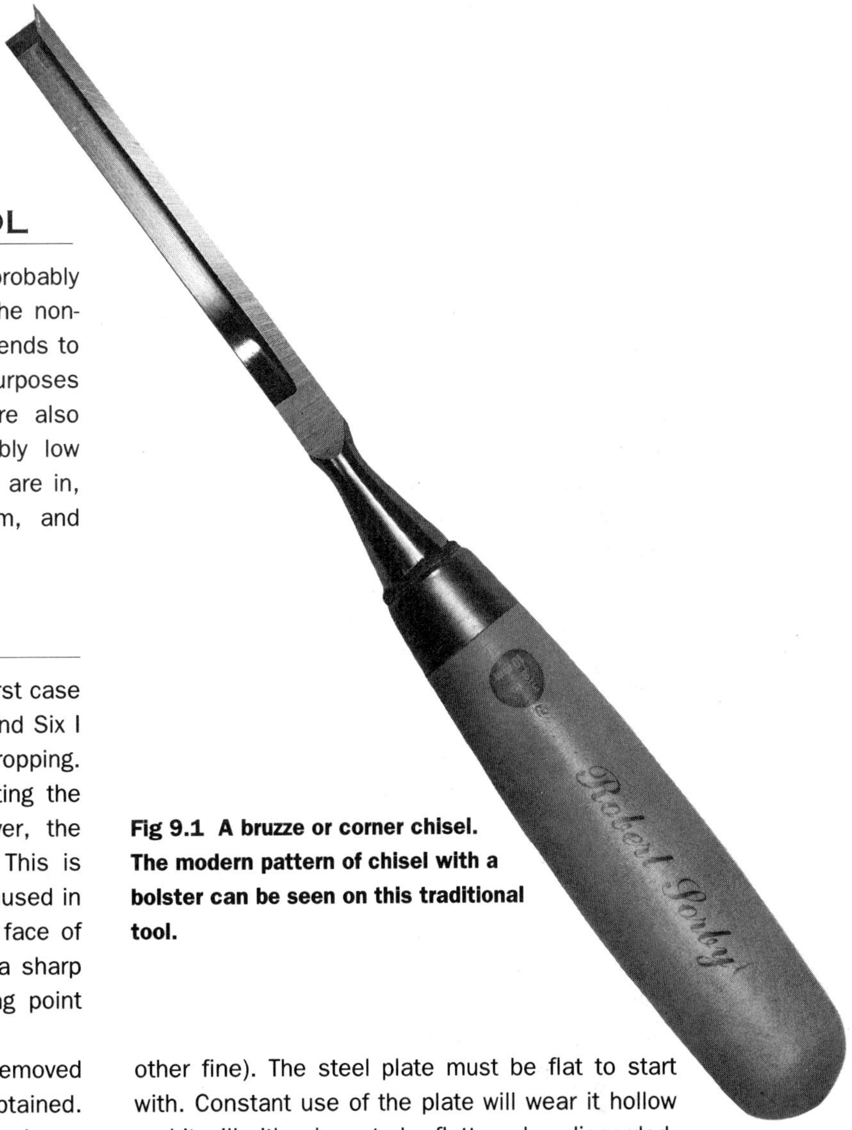

A MUCH MISUSED TOOL

Of all the woodworking tools, the chisel is probably the one that gets misused the most by the non-woodworker. One is often asked by DIY friends to sharpen a tool that has been used for purposes for which it was not intended. There are also second-hand chisels offered at remarkably low prices because of the poor condition they are in, when an hour's work can restore them, and produce a very useful tool.

DEALING WITH RUST

First let us look at how to deal with the worst case of neglect. In Chapters Three, Four, Five and Six I have discussed grinding, honing and stropping. These are the principal methods of treating the edge of the tool. There is also, however, the problem of rust pitting to contend with. This is often found on an old tool that has lain unused in some damp place for a long time. If the face of the tool is pitted, it is impossible to get a sharp edge. This can be a very good bargaining point when buying old tools.

The rust-damaged surface must be removed and a new, smooth, clean surface obtained. Rubbing the tool flat on a steel plate, using a mixture of Carborundum paste mixed with thin oil is the best way of achieving the required surface. The paste sold for grinding the valves in car engines is ideal. It is usually sold in a double-ended tin (one end has coarse paste and the

Fig 9.1 A bruzze or corner chisel. The modern pattern of chisel with a bolster can be seen on this traditional tool.

other fine). The steel plate must be flat to start with. Constant use of the plate will wear it hollow and it will either have to be flattened or discarded. Recently I have used a diamond-impregnated sharpening plate for this purpose. Rust on any other part of the tool than the face is unimportant. A clean up with some emery cloth will give the tool a presentable appearance.

JUST LIKE NEW

Once the tool's face is clear of all signs of rust, the dull surface left by the grinding paste must be removed. At this stage the face is in the same condition as that of a new tool. It is covered by a mass of tiny scratches left by the grinding medium. These scratches must be polished away by rubbing on progressively finer stones. It is particularly important that the flat surface next to the cutting edge is smooth and shiny. The rest of the surfaces are not so important and a few scratches here will not matter. Continued use and sharpening – as and when the tool needs it – will bring the whole surface to perfection. A tool in regular use that is looked after properly seems to improve with time. Sometimes a new tool, particularly a paring chisel, will be found to have a twisted face. This can be flattened using the method described above.

WHAT SORT OF EDGE?

The type of chisel and the work it is asked to do affects the grinding and honing angles. A paring chisel (which should never be hit with a mallet) works best with a very low angle. This low angle, and a sharp edge, allow the tool to cut with minimum effort from the user. Once undue force has to be used to propel the tool into the wood it will take the path of least resistance, and the user has little control over the cut being made. If the tool is correct for the job it will do as it is told.

Conversely, a tool such as a mortise chisel that is being driven into the wood with a mallet needs a strong robust edge. A delicate low-angled edge would soon break up, making the tool difficult to use. Some tools are made from tough steel which, while not taking a superfine edge, will retain a reasonable one, even when used robustly. One gets to know the different characteristics of regularly used tools. The edge best suited to each tool becomes established, and one is then confident that it is performing at its best.

THE DIFFERENT CHISELS

Over the years the names of tools have tended to change, and not everyone gives the tool the same name. I must therefore describe the tools to prevent misunderstanding. The term 'bevelled edge' never used to be applied to a particular variety of chisel. There was a bevelled-edge firmer chisel and a bevelled-edge paring chisel, but these were two quite different types. Unfortunately today the bevelled-edge firmer chisel is just called a bevelled-edge chisel and a square-sectioned, parallel-bladed chisel is known as a firmer chisel.

The Firmer Chisel

That general workhorse, the firmer chisel, is asked to perform many tasks, some of which it is unsuitable for. If a tool is expected to do a wide range of work, in the chisel's case, from paring to chopping, it will be impossible to put the ideal edge on it. The sharpening will be a compromise. This is not conducive to the production of quality work. Not everyone is in the fortunate position of having a wide selection of chisels. However, by keeping certain tools for particular types of work it is possible to get round the problem. The firmer chisel is a robust tool and is intended for medium or heavy work. It is often driven with a mallet, and the edge needs to be strong. This tool should not be ground or sharpened at a low angle; 25° should be considered the lowest grinding angle. A 27° or 28° grinding angle with a honing angle of 32° would be a good starting point.

The Bevelled-edge Chisel

This is a lighter tool than the firmer chisel and is used for more delicate work. Modern bevelled-edge chisels are thick and heavy when compared with ones made some years ago. To accommodate the requirements of modern mass production techniques, the chisel is now invariably made with a bolster where the tang meets the blade (*see* Fig 9.1), making it out of balance and unwieldy. It is far more suited to heavy work than the original light-sectioned bevelled-edge firmer chisel, which is the tool being discussed here.

The mallet is seldom used on this tool and it is used for work of a light nature. This requires a low sharpening angle; the tool can be ground as low as 20° and honed at 30°. If the honing bevel is kept narrow by frequent grinding, the tool can be used for quite delicate work. Of course, the type of wood being worked has some bearing on this. If sharpness is to be retained for any length of time when working a hard-natured wood, the edge needs to be stout.

Paring Chisels

These chisels have long thin blades and should never be struck with a mallet. Both square-sectioned and bevelled-edge blades are made. The choice of either of these is one of personal preference. Paring chisels are used for delicate paring and shaping work, and the edge needs to be kept very sharp. For most work on mild timber the edge can be honed as low as 25°. This is ideal for work such as pattern making in yellow pine, but for harder woods such as oak angles as

Fig 9.2 Sharpened tip of mortise chisel showing round at top of ground bevel.

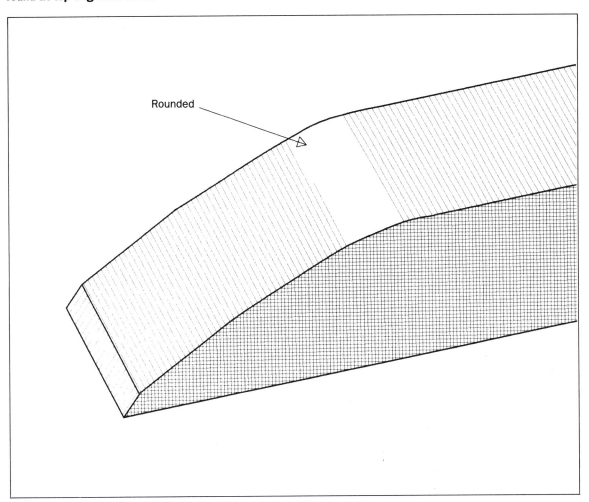

Rounded

steep as 30° may be needed. The stropping technique (described in Chapter Six) is very useful for keeping a razor-like edge on this tool.

Mortise Chisels

In the mortise chisel we have a tool that is not only designed to be driven into the wood with a mallet, but also used to lever the chips from the mortise. There are several different designs of mortise chisel and they are all very robust. The heaviest is the joiner's mortise chisel; this has an oval bolster and stout, oval handle. The sash mortise chisel is lighter and is used for lighter work. There is also a socket mortise chisel where the handle is secured to the blade by a socket instead of the more usual tang.

All these different patterns are sharpened in a similar way. The big difference between the edge of the mortise chisel and that of other chisels is the way the grinding angle is rounded off at its uppermost end (see Fig 9.2). When the grinding is left flat, the sharp corner makes levering chips from the mortise difficult. It also damages the ends of the mortise which it presses against when being used as a lever. The edge is honed from 30° to 40° depending on the type of work undertaken. Unless a steep angle is used the tool very quickly becomes blunt.

The Sash Pocket Chisel

The sash pocket chisel is one of several chisels that have particular uses. It is used for cutting the ends of the pockets in the pulley stiles of sash window linings. The wide blade is very thin and sharpened from both sides. The cutting angle needs to be as small as 20° because the cut it makes must be almost invisible.

The Drawer Lock Chisel

This tool is best sharpened with a slip stone. It can be honed on a bench stone but this is very awkward. I find the easiest method is to put a block of wood in the vice, with its upper end well above the surface of the bench. The drawer lock chisel is held on the top surface of this block to keep it steady while the edge is honed with the slip stone. This tool needs handling with some care when it is being sharpened. While concentrating on honing one cutting edge, the other one is easily forgotten and inflicts a nasty cut. To prevent this happening, wrap the part not being sharpened in a cloth. This also makes the tool much easier to hold. A honing bevel of 30° is appropriate on this tool.

The Bruzze or Corner Chisel

Tools with blades that are hollow shaped in section and sharpened from the inside are not easy to sharpen. The bruzze (see Fig 9.1) is a tool in this group. It is almost impossible to grind the bevel and all work is best done with slip stones. Holding the bruzze against the edge of the bench to keep it steady, a square-edged slip is used flat against the bevel. I have seen this tool sharpened by rubbing it along the corner of a bench stone but I have never had any success using this method myself. The problem is that the cutting edges must be kept at 90° to the length of the tool. One edge must not become longer than the other. The bruzze requires time and patience to sharpen satisfactorily, and this is where the recently introduced diamond sharpening systems are useful.

Swan-necked Chisels

Some tools have an edge that, despite being flat, cannot be sharpened on a bench stone because another part of the tool is in the way. The swan-necked chisel is an example of this. The flat side of a slip stone should be used, resting the tool to keep it steady and moving the slip. Care is needed to keep the bevel flat, as it is very easy to get a rounded edge.

Other Chisels

Occasionally other chisels will be encountered. To sharpen these tools, follow the instructions given above. For instance, some craftsmen use a pair of splay-ended chisels for cutting lapped dovetails. Apart from the fact that the cutting edge is not at 90° to the length of the chisel, there is no difference in the sharpening to that of a bevelled-edge chisel. While on the subject of dovetails and chisels, some craftsmen alter the bevelled-edge chisel so that it will fit into the corners of the lapped dovetail socket by extending the bevel down until it intersects with the flat face of the chisel. If you do this, don't allow the intersection point of the two faces to become a sharp edge. You have to hold the tool here to use it, and the last thing you need is a cutting edge.

Sharpening Gouges

Out-cannel Gouges
Any gouge sharpened on the outside can be sharpened on a bench stone. The process is similar to sharpening a chisel, but the tool is rolled sideways as it moves along the stone. Some attention is needed to ensure the cutting edge is kept square. It is very easy to wear a hollow up the centre of the stone and particular care needs to be taken to use the whole surface of the stone. Some craftsmen keep stones especially for sharpening gouges, so there is no need to worry about them being hollow; a hollow stone is better than a flat one when sharpening gouges. As with chisels, the angle the tool is sharpened at will be dependent on the quality of the steel and the purpose for which it is to be used. Although the tool is sharpened on the outside, some work is needed on the inside with a slip stone. Keep the inside flat while working with a slip to get rid of any grinding marks or edge burr thrown up while honing the outside bevel.

In-cannel Gouges
Getting a really good cutting edge on a gouge that is sharpened from the inside takes time and practice. The tool can be ground on a machine fitted with a gouge cone (see Fig 3.1). When

Fig 9.3 Honing the inside of a gouge.

sharpening gouges, the cutting edge must be kept straight and at 90° to the length of the tool. This is quite difficult with in-cannel tools. Once the edge becomes deformed it can take a lot of work with a slip stone to rectify the fault. Paring gouges with a large sweep are particularly vulnerable to this problem and extra special care is needed when sharpening them.

The tool should be held steady against the side of the bench and the top inclined away from you, so that the bevel can easily be seen. With a slip stone hone the inside of the gouge until there is no sign of the candle (*see* Fig 9.3). The outside can be polished by rubbing it on a fine stone. Rubbing the outside needs some practice to get the action right. As the gouge is rubbed down the stone it is rolled so that the whole of the surface is stoned equally. A shaped strop is very useful when a superfine edge is required.

Tool Care

Chisels and gouges when not in use are best kept in a rack (*see* Fig 9.4). If they are to be stored in a drawer, some fitment is needed to make sure that their edges are protected from knocking against anything that will cause them damage. New chisels usually have plastic edge guards on them. These are fine but somehow they seem to disappear. Probably they get dropped and cleared away in the shavings. A tool roll is the best method of protection for the craftsman who needs to transport tools about. When using a chisel be careful how it is laid down. Try to put the edge where it won't be knocked by other tools. A strip of wood at the back of the bench on which the cutting end of the blade can be placed is a very good method, but it needs discipline to place the tool correctly every time it is laid down.

Fig 9.4 Chisels in rack at the back of my bench. This is an ideal way of storing chisels: out of harm's way, yet immediately to hand when needed.

CHAPTER 10

SHARPENING CARVING AND TURNING TOOLS

SHARPENING CARVING TOOLS

It is impossible to produce quality carving unless the tools are supremely sharp. A larger proportion of the carver's time is spent in sharpening than any other woodworker's, and an edge that would be deemed acceptable by the carpenter would be considered blunt by the carver. I was taught carving by an old craftsman, who learnt his craft around the turn of the century. He told me that the craftsmen he learnt from firmly believed that if a tool became blunt the devil would sit on the edge. Never again could the tool be given a really sharp edge. While this is no doubt an old folk's tale, it is well worth bearing it in mind where carving tools are concerned. Once they become out of condition a great deal of work is required to restore them. This also applies to new tools. No carver likes a new tool. This is why when a skilled professional carver retires his tools can be sold for a higher price than new ones. Some manufacturers state that their tools are sharp and ready to use; don't you believe it.

Sooner or later we all have to put a new tool to work. The new tool from the supplier will have been ground but not sharpened. If you ever have to grind a carving tool it is usual to have the wheel turning away from you. This is the opposite of normal woodworking practice. It is much easier to control the delicate tool when it is trailing on the wheel. Most carving tools have a bevel on both

Fig 10.1 Carving gouges: lower one has normal flat bevel; upper one has the bevel rounded off where it meets the back of the blade.

faces, usually in the proportion of one-third inside, two-thirds outside. This flies against all the principles applied to most other woodcutting tools. Added to this, some carvers take the heel off the outer bevel, which is to say it is rounded off (*see* Fig 10.1), allowing the tool to follow round a hollow curve more easily. The sweep of the cutting edge is of a smaller diameter than the outside shape of the tool. The inside bevel increases the size of the sweep that is cut. This facilitates easier manipulation of the tool. Some carvers grind the corners off chisels and gouges (*see* Figs 10.2 and 10.3). Carvers who work solely on softwood (e.g. for pattern making in yellow pine) only use an outside bevel. This is the sole exception to the carver's practice of using tools with a bevel on each side of the cutting edge.

CHISELS

Carving chisels differ from ordinary woodworking chisels as both sides of the chisel are bevelled. These sharpening bevels are kept flat; they do not have grinding and honing bevels. The heel of the bevel is rounded by rocking the tool on the stone once the edge has been sharpened (*see* Fig 10.4). It is important to remember that this rounding only has the effect of removing the heel of the

sharpening bevel. The main part of the bevel remains flat. Any tendency to put a round on the cutting edge must be avoided as a round cutting edge will not allow the tool to work properly (*see* Fig 7.1). For most sharpening only a fine-grained stone is required unless the edge has been damaged, in which case a coarse stone may be used to restore the working edge. Lay the tool with the bevel flat on the stone (about 15°), with the handle in the palm of the right hand. With two fingers of the left hand, press on the blade close to the end of the blade and as near the stone as possible. Begin rubbing the tool across the stone from end to end. Take care not to rock the left hand up and down, but keep it as level as possible. Press down with the fingers of the left hand so keeping the edge pressed firmly against

Fig 10.2 Cutting edge of carving chisel.

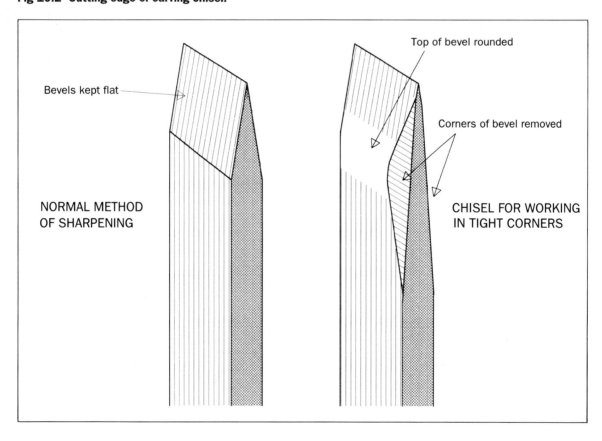

Bevels kept flat

Top of bevel rounded

Corners of bevel removed

NORMAL METHOD
OF SHARPENING

CHISEL FOR WORKING
IN TIGHT CORNERS

the stone. Practice alone familiarizes the muscles of the wrists with the proper motion. It is important to acquire this in order to form a correct habit.

When one side of the chisel has been rubbed until the whole bevel shows signs of contact with the stone, the tool is turned over and the other bevel given the same treatment. When the edge is considered the best obtainable from the stone the tool is stropped. The strop is a piece of leather dressed with a fine abrasive (*see* Chapter Five). First one side of the tool and then the other is drawn along the strop. Draw the chisel along rather than pushing it, or you will cut the strop. When the bevel has a high shine it has been stropped sufficiently. In use the tool may be

Fig 10.3 Cutting edge of carving gouge.

Fig 10.4 Putting the round on the top of the bevel of a carving chisel. The tool handle is moved up and down while the tool is rubbed back and forth on the stone. This rounds off the top of the bevel. Care must be taken not to round the cutting edge.

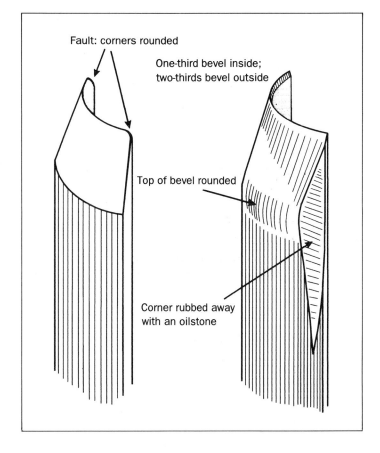

Fault: corners rounded

One-third bevel inside;
two-thirds bevel outside

Top of bevel rounded

Corner rubbed away
with an oilstone

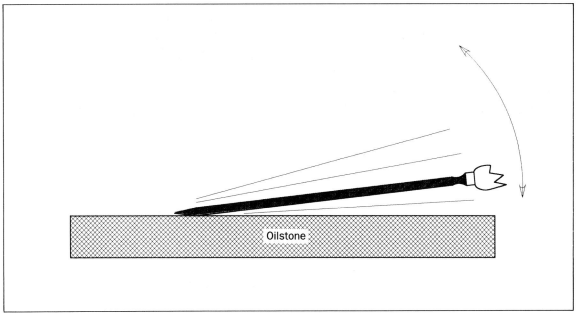

Oilstone

Fig 10.5 Honing an out-cannel gouge lengthwise on the stone.

Fig 10.6 Honing an out-cannel gouge at right angles to the stone.

stropped half a dozen times or so before it needs to be put on the oilstone again.

Skew chisels are sharpened in much the same way as the square variety. Where the tool is to be used to clean up acute internal angles, the edges of the chisel are stoned off (*see* Fig 10.2).

GOUGES

There are two ways of rubbing the outside bevel of a gouge on the stone. One way is to hold the tool in much the same way as described for the chisel, but roll it as it progresses along the stone so that the whole bevel is sharpened. Alternatively the tool can be held at right angles to the stone, and then rolled as it passes down the stone (*see* Figs 10.5, 10.6 and 10.7). In this way every part of the bevel is rubbed evenly on the stone. Be careful to keep the edge of the tool straight as it is very easy to remove the ears (outermost corners of the cutting edge) from the tool, and rob it of much of its usefulness.

There is a tendency when sharpening gouges for the stone to become misshapen. It wears hollow in its width, making it useless for sharpening flat tools. To prevent this, some carvers use both sides

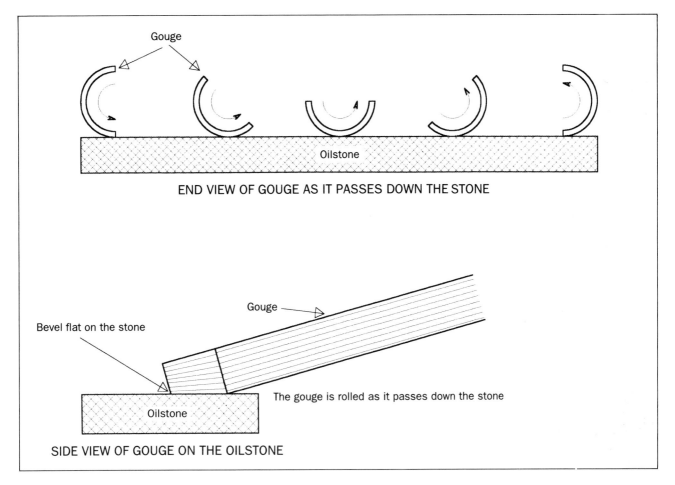

Gouge

END VIEW OF GOUGE AS IT PASSES DOWN THE STONE

Oilstone

Gouge

Bevel flat on the stone

The gouge is rolled as it passes down the stone

Oilstone

SIDE VIEW OF GOUGE ON THE OILSTONE

Fig 10.7 The rolling motion of the gouge as it is honed.

of the stone. They reserve one side for gouges and the other for flat tools. If the heel of the bevel is to be removed, the angle the tool is held at can be reduced and the tool given a few passes down the stone.

The inside bevel is sharpened with a slip stone of suitable profile. Professional carvers have many oil slip stones. Their shapes are modified to fit the tools when a shop-bought slip of the correct contour is unavailable. The novice carver by careful manipulation can make do with about half a dozen. The secret is to ensure that the whole cutting edge gets equal treatment. There is

nothing worse than trying to work with a tool that does not have a straight cutting edge. It is surprising how quickly this fault can develop if constant attention is not paid when stoning the inside bevel. This fault easily occurs when using a slip of smaller radius than that of the sweep of the gouge.

To strop the outside bevel of the gouge it can be drawn over the flat strop rolling it from one side to the other as it progresses. When it comes to the inside there is a choice. A piece of sheet leather can be bent so that the bend forms a radius that approximates the inside shape of the gouge. Alternatively, leather can be glued around a shaped block (*see* Fig 10.8). While the inside is being stropped the tool is held against the edge of the bench to steady it. Strops need to be kept clean. The greasy nature of the strop dressing picks up grit and dust very easily. Any of this debris on the strop can spoil the edge of a tool. A small box in which the collection of strops can be kept does not take long to make and is a worthwhile project.

VEINERS

The veiner requires similar treatment to the gouge. As this tool's section does not form part of a

Fig 10.8 Leather glued around a block of wood to make a shaped strop.

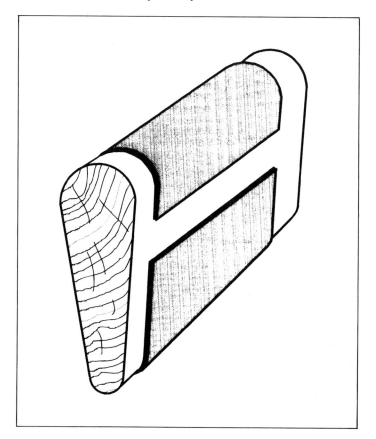

circle – it has straight sides – only half can be honed at a time. It is as well to give the straight side one stroke or so in every half dozen all to itself, just to keep it in shape. Special care needs to be taken when sharpening this tool as it is easily rubbed out of shape. Stropping the inside of the veiner presents some difficulty. It has to be sharpened as three separate parts; the two straight sides and the rounded bottom. It also needs a very narrow slip; a knife edge slip may be required inside the straight sides. The edge of a stiff, thick piece of leather can be shaped to fit inside small veiners. Some craftsmen use a strip like a leather boot lace, which they draw down the cannel and over the edge of the tool.

THE PARTING TOOL

On first acquaintance the parting tool looks so simple that one would suppose it an easy tool to sharpen. After all, it is just as if two chisels were joined together at an angle. However, this tool is a tricky customer to sharpen properly. It is the point where the two blades join that presents the problem. The outsides can be stoned exactly like two chisels. First one side, then the other, taking particular care not to make one side longer than the other, and keeping each side square to the apex. For the inside, a knife edge slip stone is required to put the inside bevel on each side. Now we come to the tricky bit. What has happened at the point in the centre where the two straight parts of the tool meet? The tool has a slight radius on the inside, but the two outside straight faces meet in a sharp point. The consequence of this is that the tool is thicker at the apex than it is along the two straight sides. This causes a beak to be formed (*see* Fig 10.9) or, if one side has been rubbed more than the other, a hollow has appeared. This is because the outside faces are meeting off-centre and the inside vee is now coming out of one of the sides. To get rid of the hollow, the outside must be rubbed to correct the inaccuracy. To remove the beak, round the outside

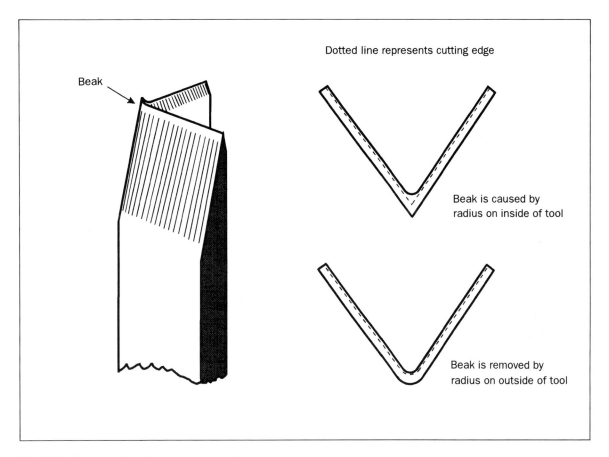

Dotted line represents cutting edge

Beak

Beak is caused by
radius on inside of tool

Beak is removed by
radius on outside of tool

Fig 10.9 How a beak is formed on a parting tool.

where the two sides meet. Thus the sharp angle becomes a gentle round and the whole of the cutting edge is straight.

CURVED, BENT TOOLS & SPOONS

The special-shaped tools, whether they be front bent, back bent, or curved, require similar treatment to the straight tools. The spoons and front-bent gouges present a small problem in that the slip fouls on the curved part of the tool. Because of this, only the end of the slip can be used. The secret is to take your time, and do not round the edge over. When stoning the outside of spoon tools they need to be held at a much higher angle than is customary with straight tools. If the tool is presented to the stone and the handle raised until the bevel is flat on the stone, the correct sharpening position will be established. Particular care should be taken to ensure that all the bevel gets the same amount of rubbing. Some people find it easier to sharpen one half of the tool first, before reversing it and sharpening the other half.

MACARONIS, FLUTERONIES & BACKERONIES

This group of tools, while having exotic names, are sharpened according to the same techniques as those described for other tools. Once the basic techniques are grasped, it is a matter of taking one's time, ensuring that the edge configuration is kept correct and the tool is never allowed to get out of condition.

SHARPENING TURNING TOOLS

I have repeated the question several times in this book, and I make no apology for repeating it again here: 'How sharp is sharp?' When it comes to the subject of turning tools, there are so many opinions about what constitutes a cutting edge that it is difficult to know where to start. Some turners only use a grinder to sharpen their tools; an oilstone is never resorted to. Others not only use the oilstone but also strop their tools. Between these two extremes there are many alternatives. In my opinion a tool that has a really keen edge leaves a much better finish on the work than a tool straight from the grinder.

Before you decide how you will sharpen your turning tools, make sure you understand just what each tool should do, and how it does it. Try different methods of sharpening before finally deciding what suits you best. Give each a good try out on several different timbers. A tool that is working to perfection can be very pleasurable to use.

DIFFERENT TOOLS & MATERIALS

During my time in the trade I have seen so many different turning tools introduced on to the market that I wonder how we ever managed before. The craftsman who taught me in the late 1940s had no more than half a dozen tools, and only three of these were in regular use. This gentleman was paid piece work, and his wage was completely dependent on the amount of work he produced. Even so he spent a good proportion of his time sharpening tools. He taught me that the accuracy and finish of his work were dependent on the sharpness of his tools.

Turning tools have traditionally been made from high-carbon steel, but in recent years tools made from high-speed steel have been available. This material holds a working edge much longer than high-carbon steel. There are also tungsten carbide-tipped tools. I have used the latter for turning glass-reinforced plastic mouldings, but have not tried them on wood. Apart from these different materials from which the tools are made, many new shapes of tool are available, but there is a danger here for the novice. Proficiency in the use of a tool comes through familiarity with it. If one keeps changing tools, complete mastery is never achieved. Even the shape of the cutting edge of a tool varies from one user to another. Take the spindle gouge for instance. One craftsman sharpens it square to the length of the tool, whereas another will sharpen it so it is shaped like a fingernail. The beginner is in a quandary here, unsure of which approach to take. Unfortunately it is difficult to offer advice; all that can be done is to try everything, and then adopt whatever works best. Even so, as one's ability increases, more changes will be necessary.

CONTROVERSY

Woodturning is surrounded by more controversy over tools and sharpening than any other branch of woodwork. I am old-fashioned and like to use the traditional methods. However, woodturning has become a popular hobby, and sharpening takes up precious time that hobbyists would prefer to spend at the lathe. Therefore mechanical means of sharpening have been introduced and adopted by many amateurs. There are three different ways of sharpening turning tools:

■ Sharpening on a dry high-speed grinder only.
■ Grinding on a machine and sharpening on an oilstone.
■ Grinding on a machine and buffing the edge by machine.

Each of these methods has its devotees. The second is the traditional method, and is used by most time-served professionals. The finish of the cutting edge affects the finish of the workpiece. There should be very little need to use abrasive paper if the tools are sharp and used properly. In the third method a rubber abrasive wheel can be used in place of the oilstone, or the final edge can be polished on a fabric wheel dressed with an abrasive compound. If machine sharpening is used for turning tools a flat bevel must be maintained. A rounded bevel will not allow the tool to be presented to the work at the correct cutting angle.

Turning tools can be divided into two distinct groups: tools used for cutting and tools used for scraping. The scraping tools are not used by all turners. There are arguments as to whether or not they leave a finished surface. I have seen turners who get such a high finish from the cutting tools that scraping is unnecessary, and in fact would spoil the work. However, this degree of skill takes much practice and experience to acquire. The scraper is an easy tool to use and is often the only way a beginner can get a finish on the surface of the work. This is the thinking behind many new tools; their shape has been developed so that the technique of using them is easier to master. Turning tools come in a variety of sizes, the largest being called 'long and strong'; normal sized tools are called 'standard' and the smallest are called 'miniature'.

GOUGES

Roughing-out gouges are normally of the long and strong variety, with a deep fluted blade. They are sharpened square with a bevel of 40° to 45°. The bevel on all woodturning tools must be flat, not rounded or hollow. There are no separate grinding and honing bevels. Some people find it difficult to get a flat ground bevel on the periphery of a grinding wheel (*see* Chapter Three for a description of grinding techniques). There

Fig 10.10 Honing a turning gouge; the tool is rubbed on a stationary stone.

may be a need for a machine that allows grinding on the side of the wheel. I prefer a water-cooled wheel, but many turners use a standard high-speed bench grinder. There are several grinding machines on the market that have belt-grinding attachments, which produce a flat bevel. It is possible to turn a wooden disk in the lathe, and glue a sheet of aluminium oxide paper to the face to make a flat grinding surface. The only problem is that it has to be mounted in the lathe when it is to be used, and this requires the removal of the workpiece.

There are three ways of honing the bevel. Firstly, the gouge can be held still and the stone rubbed on to the bevel. Secondly, the gouge can

Fig 10.11 Sharpening a turning gouge on a high-speed grinder.

be rubbed on the stone which is placed on the bench. The gouge is kept longways on to the stone and is rolled from side to side as it passes along the stone (*see* Fig 10.10). The third method is to place the stone on a flat surface and rub the gouge at 90° to its length. After stoning the bevel, any burr is removed with a slip stone (*see* Fig 10.12). When using the slip stone on the inside, take care not to introduce an inside bevel. The inside flute of the gouge must be kept straight and flat.

The spindle gouge is a shallow tool. Most turners make the cutting edge the shape of a lady's fingernail. It is sharpened by using the same technique as that used for the roughing-out gouge, although the bevel angle is often less than it is on that tool. You may find it interesting to experiment with this angle, but do not go below 30°. As with other cutting tools, the steeper the bevel the stronger the edge. There may be an argument for steeper bevels on hardwood than on softwood.

In recent years many bowl gouges have been made from a round bar with a flute machined in it. These are sharpened in the same way as a conventional bowl gouge, with the edge kept square to the length of the tool. A bevel angle of 45° is usual. However, when working on deep bowls this is often increased to as much as 60° to ensure that the bevel can be kept in contact with the work piece.

It is difficult hollowing out end grain, say, when making an egg cup. A ring tool is useful for this task. These are usually made from high-speed steel. Sharpening is not difficult; a diamond file (*see* Chapter Five) is the best tool to use. Make sure all the original cutting angles are maintained.

CHISELS

Turning chisels have either square or skew-cutting edges. When it comes to sharpening there is no difference in the methods used. The bevels should be kept the same on both sides of the tool. An included angle of 30° is normal. The sharper the tool the easier it is to control. I find that after the tool has been ground and honed, a few passes along the surface of a Japanese polishing stone gives it that little extra something. Here again, it is essential that the bevels are flat. Do not introduce a secondary honing bevel.

PARTING TOOLS

A parting tool can be likened to a narrow chisel with a thick blade. When you come to sharpen it, treat it just like a chisel. A recent introduction is a parting tool that has a flute along one edge and a single bevel. The single bevel is ground and honed. The flute along the edge will produce a cutting edge that has two little ears on it. This tool leaves a very smooth face on end grain.

The length of time that a turning tool can be used before it requires sharpening is much less than the beginner realizes. Think of the amount of wood the turning tool cuts in a given time. Compare this with the amount of wood cut in the same time by a hand tool used at the bench. Using a chisel at the bench, even on a mild timber, one would expect the edge to last no longer than twenty minutes. A skew chisel on the lathe will cut a similar amount of wood in less than five minutes. There is another point worth considering: if a tool is allowed to rub against the revolving work in the lathe without taking a proper cut, the edge will immediately be lost. A tool scraping the surface loses its edge much faster than a tool taking a thick shaving.

SCRAPERS

The cutting edge of the scraper can have many different shapes. The tool is used straight from the grinder, and a burr left by the grinder is in fact the cutting edge. The bevel on the scraper is ground at 80°. Whatever contour the end of the scraper is shaped, keep it smooth. Any irregularity will be reproduced on the workpiece.

CARE OF THE TOOLS

Most carvers use a tool roll to keep their carving tools in when they are not in use. I find a nest of shallow drawers better. The drawers with the tools in them can be put on the bench. The tools are used and replaced in the drawer as the work progresses. The only problem is chips of wood from the carving get in the drawers, and they need clearing out from time to time. An occasional wipe over with a rag moistened with a thin oil prevents the tools from rusting.

Fig 10.12 Removing the burr from the inside of a turning gouge.

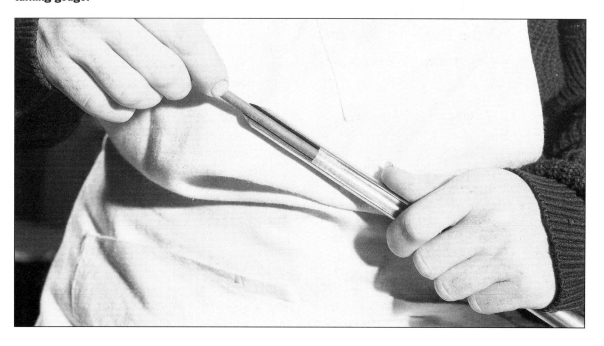

CHAPTER 11

Sharpening Saws

Tooth Configuration

There is no mystique about sharpening a saw. If the job is approached logically, with a knowledge of what is required, anyone can master the technique. First we must understand how the saw cuts the wood. The shape of a saw's teeth varies according to the type of work it is used for. A saw for ripping down the grain has different teeth to that of a saw used for sawing across the grain. The tooth configuration of a saw used for both ripping and crosscutting must be a compromise. Think of the wood as a bundle of straw with all the individual straws parallel to one another. It can be seen that cutting along the straws is very different from cutting across them. In the first case, straws are removed from each other, and in the second case they are severed. Sawing wood is very similar to this, with the straws replaced by wood fibres. The teeth of the saw used for ripping can be likened to a row of little chisels, while the crosscuts resemble a row of knives (*see* Fig 11.1).

Nearly all saws are sharpened with a triangular file with an equilateral triangular section. That is, all the sides are the same length, so that the angle at each corner of the file is 60°. The leading edge of one tooth and the back of the preceding one are filed at the same time. Therefore, the shape of the teeth is governed by the shape of the file. The teeth of a crosscut saw are more upright than those of the ripsaw. That is to say, the ripsaw's teeth lean forward. The angle the front of the tooth has to a line drawn through the tips of the teeth is known as the **rake angle** (*see* Fig 11.2). The rake of ripsaw teeth is normally 90°, while those on a crosscut are 75° to 80°.

The Set

The slot that the saw cuts is known as the **kerf**. This must be wider than the saw blade's thickness, otherwise, as the saw cuts it will bind in the kerf. To make the saw cut wider than the gauge of the blade the teeth are bent out at their tips; this is known as the **set**. Energy must be expended to remove wood from the cut as sawdust. The wider the cut the more energy required. Sawing requires less energy if the kerf is as narrow as possible. Therefore, the minimum of set is used. As the tooth has to be bent, care must be taken not to break it out of the saw. Consequently it is important that only the top third of the tooth is bent. There is a weak point across the bottom of the tooth on the gullet line, caused by the sharp angle in the bottom of the gullets. Saw files do not have sharp corners; their sides meet in a small radius. This makes the tooth stronger, as the bottom of the gullet has a small radius and not a sharp corner (*see* Fig 11.3). Even so, only the tip of the tooth is set. Having read this far it would be a good idea if you looked closely at a few saws. Inspect their teeth. You can probably identify the parts I have described.

Perhaps the saw you are looking at has inductance-hardened teeth. These are a dark blue colour that stands out against the bright, shiny surface of the saw. If you are not sure, try to file the last tooth under the handle. If it is very hard and the file only just marks it, then the saw is inductance-hardened. These saws are made to be disposable. Once they become blunt they are discarded and a new one purchased. This may seem a sensible practice to some people, but

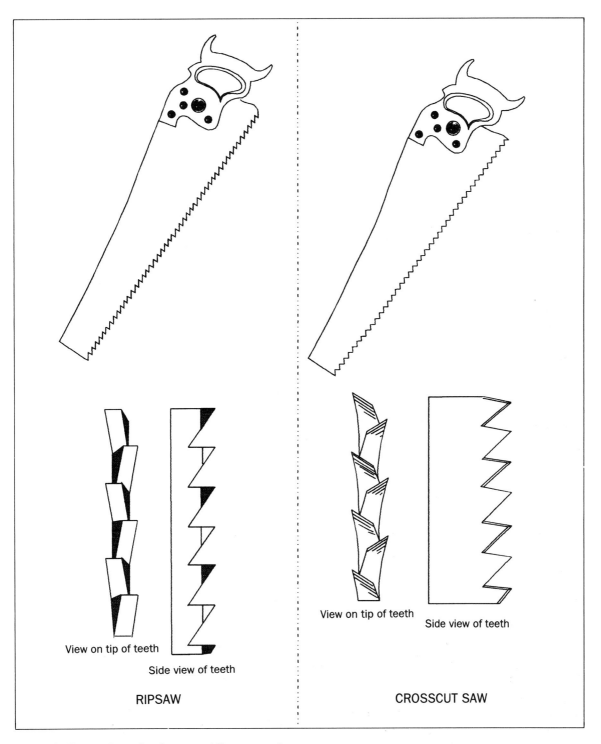

View on tip of teeth

Side view of teeth

RIPSAW

View on tip of teeth

Side view of teeth

CROSSCUT SAW

Fig 11.1 The teeth on the ripsaw and the crosscut saw compared.

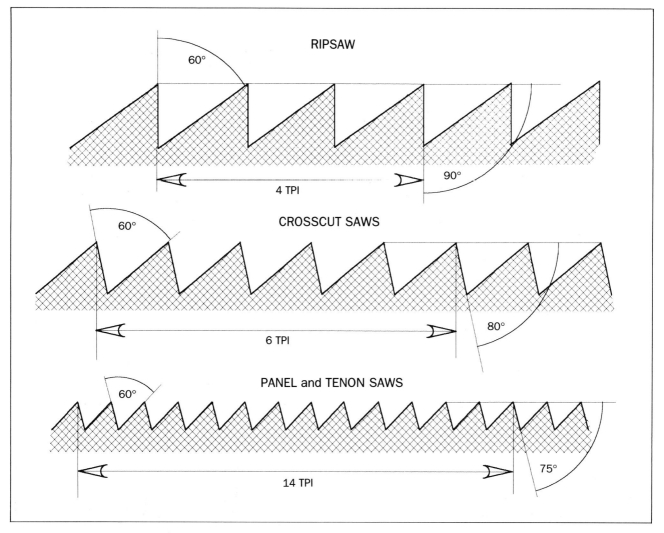

Fig 11.2 Side view of teeth on different saws showing rake angle.

there is a down side to it. Because the saw is disposable its initial cost must be kept down. Therefore, the blade is not properly balanced and taper ground. When one has used a quality saw, the difference is immediately apparent. Even saws with specially hardened teeth soon lose their very keen edge. A saw that can be kept really sharp, with teeth shaped and set for the work it is used for, is a much sounder investment than a cheap throwaway. Recently, several manufacturers have

produced saws with tooth shapes based on Japanese saw technology. All of those that I have seen have hardened teeth. This may be fine for the DIY enthusiast or for use on a building site, but not in the workshop.

TPI

Saw teeth not only vary in shape, they also vary in size. For instance, a dovetail saw may have 24

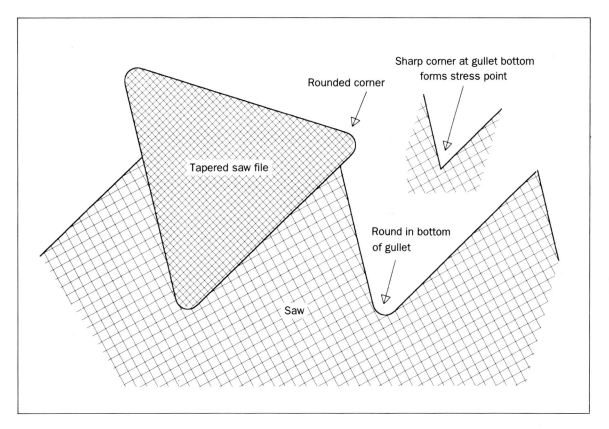

Fig 11.3 The round corner on the saw file forms a round in the bottom of the saw's gullet.

Fig 11.4 Saw file twice the size of gullet as recommended by most text books; however, this masks the tooth being sharpened.

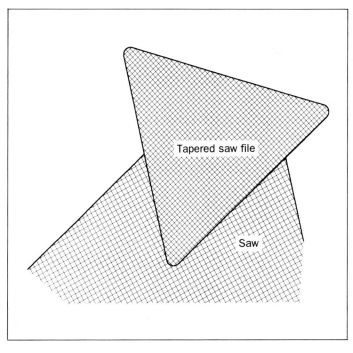

teeth per inch (TPI) while a full rip will only have 3½ TPI. TPI can be a bit confusing, since the tooth at the start and the tooth at the end of the inch are both counted (*see* Fig 11.2).

The size of the file used to sharpen the saw has a relationship to the size of the saw's teeth. The section of the file used, according to most textbooks, should be approximately twice the size of the gullet (*see* Fig 11.4). Personally I find this a bit on the large side. The reason given for this recommendation is that as each of the three corners of the file are used, new file teeth will be

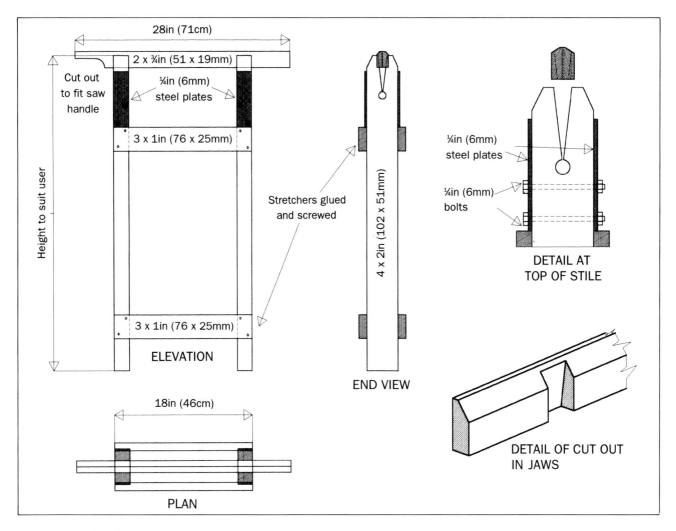

28in (71cm)

2 x ¾in (51 x 19mm)

Cut out to fit saw handle

¼in (6mm) steel plates

3 x 1in (76 x 25mm)

Height to suit user

Stretchers glued and screwed

4 x 2in (102 x 51mm)

3 x 1in (76 x 25mm)

ELEVATION

END VIEW

¼in (6mm) steel plates

¼in (6mm) bolts

DETAIL AT TOP OF STILE

18in (46cm)

PLAN

DETAIL OF CUT OUT IN JAWS

Fig 11.5 Saw sharpening block.

available. Teeth on the sides of my files never wear out; it is the corners that are the first to go. I find that a file just a little larger than the gullet suits me best. This enables me to see more of the teeth I am sharpening, whereas a larger file masks the teeth being filed.

Hold it Firm

A major requirement when sharpening a saw is that the tool is held rigid. Unless the saw is firm and steady it is impossible to file the teeth properly.

There are several different pieces of apparatus that can be used to accomplish this. Most craftsmen make a **saw sharpening block** (*see* Figs 11.5 and 11.6). This can be made to suit the height of the craftsman so that the saw is at exactly the right height when sharpening. There is nothing worse than stooping over the saw while trying to concentrate on sharpening. The teeth of the saw need to be at elbow level or a little higher. The saw block takes up precious workshop space when it is not being used. In a large workshop this is of no consequence, but where this is a

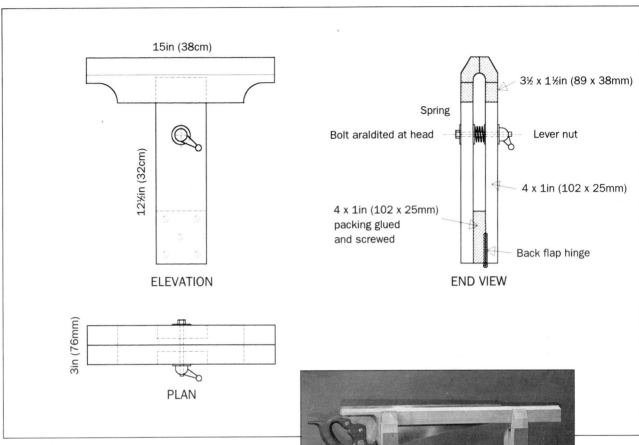

15in (38cm)

12½in (32cm)

ELEVATION

3in (76mm)

PLAN

3½ x 1½in (89 x 38mm)

Spring

Bolt araldited at head

Lever nut

4 x 1in (102 x 25mm)

4 x 1in (102 x 25mm)
packing glued
and screwed

Back flap hinge

END VIEW

Fig 11.7 Saw sharpening chops.

Fig 11.6 Saw sharpening block.

nuisance, saw chops are used instead (*see* Fig 11.7). Once saw sharpening vices were made, but unfortunately none are manufactured today, though they do appear on the second-hand market from time to time (*see* Fig 11.8). It is possible to make do by clamping the saw blade in the vice between two boards, but this is not recommended, unless you want back trouble.

When I had to teach apprentices to sharpen saws, I would offer all the local gardeners and handymen a free service. This enabled the apprentices to practise on saws that were not their own, which reminds me of a piece of advice I

Fig 11.8 Saw sharpening vice.

takes at least half an hour to sharpen a saw.

The first saw you attempt to sharpen should have large teeth, the larger the better. Secure the saw with just the teeth and a very small amount of blade above the jaws of whatever device you are using to hold the saw. You must work in very good light. If you can work so that the saw is between you and a window, this is ideal. There are five different operations encountered when sharpening a saw. I will explain each of these in the order that they are normally performed.

TOPPING

Topping is also known as **breasting** or **jointing**. There is some debate among craftsmen whether this is always necessary. It consists of running a flat mill file along the top of the teeth to bring them all to the same height. It is very easy when sharpening a saw for the teeth on one side to become slightly longer than those on the other. In use, some teeth may have come into contact with a foreign object and been damaged at the tip. Topping removes these inaccuracies and puts a small, flat surface on the tip of each tooth. This newly filed surface will catch the light and show up very clearly. It acts as a datum to which the tooth can be filed. If the saw is in very good condition and only needs a touch with the file to bring it back to sharpness, topping can be omitted. You may have trouble holding the file exactly square to the blade, in which case a topping guide can be made (*see* Fig 11.9). Fig 11.10 shows topping being carried out using a guide.

SHAPING

The purpose of this operation is to restore the teeth to their proper shape. A saw in good condition does not require this treatment. Once you have been sharpening saws for a while you will be surprised at the poor condition of many saws. The ability to restore them will enable you to take a tool that is next to useless and make it

was offered by an old craftsman when I was a lad: 'Never lend your saw to anyone. If you have to cut wood that is dirty or may have a nail in it, borrow a saw'. I once worked in a workshop that had nearly 60 craftsmen in it. There would be a constant din of hammering and sawing, etc., yet if somebody was unfortunate enough to hit a nail with a saw, that dreaded sound would bring the whole workshop to silence. The point that I am trying to make here is, be careful, not only when sharpening your saw but also when you are using it. It only takes a few seconds to clean grit from the surface of a piece of wood with a wire brush. It

A little wet grinder, ideal for the small workshop (*see* page 24).

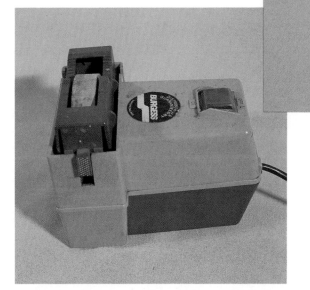

A starwheel grindstone dressing tool (*see* page 23).

A planer blade grinding jig made by the author, mounted on a pillar drill (*see* Chapter 3).

The Sharpenset grinding machine fitted with a planer blade grinding attachment made by the author (*see* page 25).

DMT diamond-coated files are colour coded, each grade being a different colour (*see* page 47).

A new Arkansas oilstone. Smith's stones are supplied in a pencil cedar box (*see* page 34).

A four-sided strop originally intended for stropping cutthroat razors. This is a very useful item for the woodworker (*see* pages 49–50).

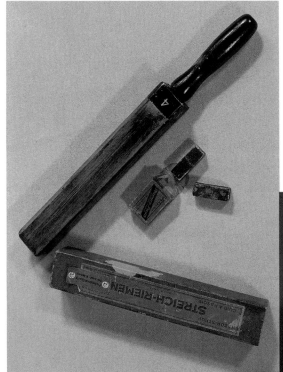

A German sharpening machine. The two felt wheels are dressed with a very fine abrasive. The texture of the felt is coarse on one wheel and fine on the other (*see* pages 54–5).

A cabinet scraper with ticketer made from an old three square file (*see* pages 124–6).

The back of two carving gouges. The one above has the bevel rounded off, while the other has a straight, flat bevel (*see* page 71).

A jig for sharpening chain saws, fitted on a machine as it would be in use (*see* pages 112–13).

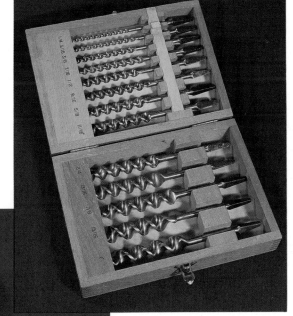

A set of Jennings pattern bits in a fitted case. This is the best way to store bits (*see* pages 97 and 104).

Working up a slurry on the face of a Japanese water stone with a *Nagura-to* (correcting stone) (*see* page 134).

Occasionally the faces of Japanese plane irons and chisels need to be rubbed back on a steel plate, using Carborundum powder or paste (*see* pages 136–7).

A double-edged Japanese saw with cover (*see* pages 144–5).

The face of a really sharp plane iron must shine like chromium plate (*see* page 7).

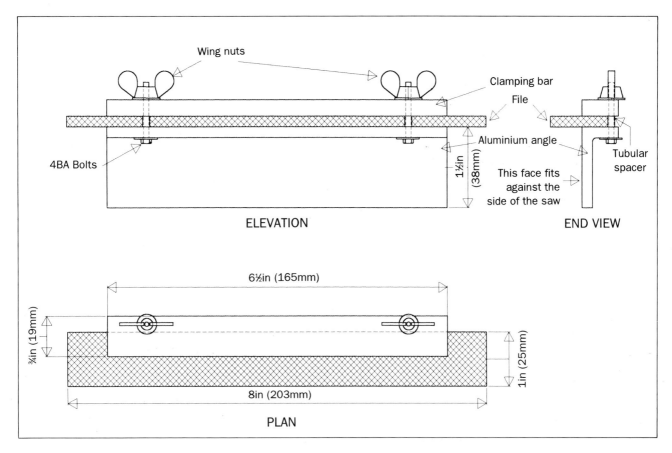

Wing nuts

Clamping bar

File

Aluminium angle

This face fits
against the
side of the saw

Tubular
spacer

4BA Bolts

1½in
(38mm)

ELEVATION

END VIEW

6½in (165mm)

¾in (19mm)

1in (25mm)

8in (203mm)

PLAN

Fig 11.9 Topping guide.

work as well, or even better than when it was new. So, if you can carry out this task it opens up a source of useful tools.

Mount the previously topped saw so that its teeth can be seen in silhouette. The misshapen teeth can now be clearly seen. Start at one end of the blade, it does not matter which, and place a saw file of suitable size, held at 90° to the blade, in the first gullet. File the gullet, taking half the topped flats from the teeth on either side. The triangular file will cut the front of one tooth and the back of the next. To make the teeth regularly shaped, pressure may have to be applied more on the front of one tooth than the back of the next. This is to say, some sideways pressure is applied to the file. If you find difficulty in keeping the pitch

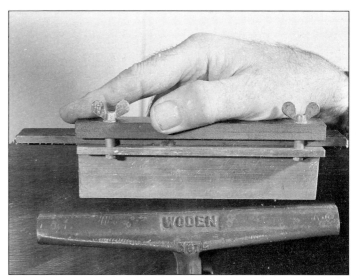

Fig 11.10 Topping a saw with tool illustrated in Fig 11.8.

of the teeth correct, try the following. A small block of softwood about ¾in square and 1½in long (19mm x 38mm) has a hole bored through its length in the centre of its section. This is forced on to the tip of the file so that when the file is held at the correct angle, the surface of the block is level (*see* Fig 11.11). If the face of the block is kept level, the pitch of the teeth will be correct.

All the gullets are filed in the same way from one side of the saw, with the file straight across the blade. The process may need repeating a couple of times to get all the teeth absolutely regular. After shaping, all the gullets should be of the same depth, and all the teeth should be the same size and at the same pitch. Hold the saw against the light and inspect it from both sides. Most saws that need shaping have irregular teeth caused by somebody trying to sharpen them without having the necessary knowledge. This produces an alternating series of large and small teeth throughout the length of the saw, known in the trade as 'cows and calves'.

SETTING

It is usual to set a saw before sharpening it, although this may not be necessary when a saw is in good condition and only requires a stroke or two

Fig 11.11 Blocks of wood applied to the saw's edge and the file's end to help maintain the correct angles when sharpening.

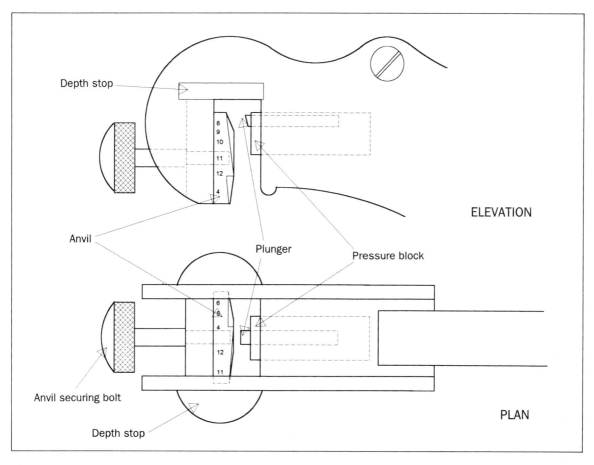

Fig 11.12 Pliers saw set

of the file on each tooth to sharpen it. If the saw is one that you use yourself, you will be aware of the width of the kerf it cuts. If this tends to be tight on the blade, setting will be required. Saws used on wet or green wood require much more set than those used for general woodwork. There are several different ways of applying the set to a saw. I would not advise you to use the anvil and hammer as a saw doctor does, as this takes a lot of skill and practice. Without doubt the easiest tool to use is the **pliers saw set**. There are several different makes on the market, all similar and based on the same principles. When buying a pair of saw-setting pliers, make sure they are made from metal. There are some very inferior plastic

ones about. The plastic bends in use and the saw gets irregular setting.

The pliers saw set has four active parts, allowing the set applied to the saw to be adjusted. The **anvil** is a thick disc of metal that has a graduated chamfer running around its face at the outer circumference. Around the edge of the anvil is a series of numbers. On most tools these are the suggested settings and relate to the TPI of the saw being set. I find the set given by this recommendation a bit on the coarse side. With a little experience the correct setting of the anvil is no problem. The pliers are placed over the tooth to be set. A **depth stop** fits on top of the teeth and positions the tool at the correct level on the saw.

As the handles of the pliers are squeezed together, a **pressure block** moves forward and pushes the blade tight against the anvil. The **plunger** then forces the tip of the tooth against the anvil, bending it to fit the chamfer. The pliers are moved to the next tooth but one and that tooth is set. This procedure is carried out for the whole length of the saw from both sides (*see* Fig 11.12).

If the saw has just had its teeth reshaped, it will be found expedient to rub the blade flat on an oilstone. This will remove the burr left by the file. A burr makes it difficult to position the saw-setting tool on the teeth. With most pliers saw sets there is no need to use much force when squeezing the handles together. Should the tip of the teeth show flattening from the plunger, too much pressure is being applied.

SHARPENING

The file used to sharpen the saw is a special tool; it is not an engineers' three square file, but is properly called a tapered saw file. These tools can be single or double-ended. For sharpening most workshop saws, the double-ended file is the easiest to use. The size of the file used is related to the size of the saw. I find a 7in (178mm) double-ended about right for my tenon and panel saws. When it comes to the hand saw (6 TPI), I use a 10in (254mm) double-ended. The full rip at 3½ TPI needs an 8in (203mm) single-ended file.

All files should be fitted with securely attached handles, partly because the sharp tang on single-ended files can inflict a nasty wound to the palm of the hand; but also because sharpening a saw requires control of the file, and without the handle the file is a difficult tool to hold exactly as required. On page 82 I discussed tooth shape, and it is important that this is maintained when filing.

The saw is positioned as it would be for shaping (*see* pages 88–90). If the file is pressed down in a gullet it will take up the angles used previously to sharpen the saw. Of course this is not so if the teeth have just been reshaped, and it is not necessarily the correct angle at which the saw should be sharpened. We now come to a couple of controversial points, both of which can cause arguments in the workshop. Is the file kept level when sharpening or is it allowed to incline upwards away from the handle? I keep it level. Which end of the saw do you start sharpening from, the tip or the handle? Personally I don't think it matters very much.

The ripsaw is sharpened with the file at right angles to the blade. This produces teeth with tops resembling the edge of a chisel. Crosscuts are sharpened with the file at an angle, which can vary from 45° to 65°. 45° is used for cutting softwood and 65° for hardwood. This is being very precise, and an angle of around 55° will be found suitable for most work. This angle must be maintained as each tooth is filed. Sometimes this can prove very difficult and some form of guide is required: a piece of wood 6in (152mm) long and 1in x ¾in (25mm x 19mm) in section with a saw cut across its width in the middle of its length (*see* Fig 11.11). The saw cut is at the sharpening angle. When the saw cut is placed over the teeth of the saw being sharpened, the block sits on the saw with its edge at the sharpening angle. The block can be moved along the edge of the saw as sharpening proceeds. A good test for a properly set and sharpened crosscut is to see if a needle will slide down the length of the saw between the teeth without falling off (*see* Fig 11.13).

The angles given above are those generally accepted in the trade for workshop saws. Other saws used in the garden for pruning and similar jobs may be different. If the saw is in reasonable condition, then maintain the angles that are already on it. Japanese saws are totally different to our Western saws and are discussed in Chapter Fourteen.

HONING

Sharpening the saw with the file leaves a burr on the sides of the teeth. This burr will not affect

Fig 11.13 A needle will glide down between the teeth of a properly sharpened and set crosscut saw without falling off.

rough sawing, but for quality work it is unacceptable. The saw should be laid flat on the bench and a fine oilstone rubbed flat along the teeth a couple of times. The saw should then be turned over and the other side treated in the same way. Not only does honing remove the burr but also any slight inaccuracy in setting is corrected. The honing also turns the tips of the crosscut – which after filing are little sharp points – into knives.

CARE OF THE SAW

Where the saw is only used in the workshop it is best hung up when not in use. The problems start when the saw has to be transported about. It has been traditional to make a **saw keep**. This is a strip of wood a little longer than the saw with a saw cut along the length of one edge. This strip is tied on to the saw when it is not in use. In recent years, new saws have been supplied with a plastic saw keep. Unfortunately these plastic keeps do not last long. The saw teeth soon cut through the plastic. Saw cases made from canvas can be bought from retail shops. Years ago craftsmen would get the local saddler to make them a case out of leather. Occasionally one of these turns up second-hand at a carboot sale. If you have the opportunity to obtain one of these, consider yourself lucky. Wipe the saw over occasionally with an oily rag, particularly if it has been used in damp wood.

CHAPTER 12

SHARPENING HOLE-BORING TOOLS

MANY TYPES

There are many different types of bit used for boring holes. The terms 'boring' and 'drilling' cause some confusion. Metalworkers and engineers use the term drilling for most hole-making processes, while woodworkers use the term boring. However, both professions use twist drills. The only time woodworkers use the term drilling is when using the twist drill. Because of the many different patterns of bit, and the way they have been adapted for different purposes, the same or similar tool may have several names. I have tried to use the name in most common use. However, in most cases the tool is also illustrated to try and avoid confusion. There are still many old shell, spoon and nose bits about, although it is many years since these have been made. As these and some other tools used in the brace seem to last forever, I will endeavour to describe the sharpening technique for them all. The technique is very similar for most, except for the twist drill. A problem encountered sometimes with an old tool that has got out of condition is determining the correct shape of the cutting edge. Make sure you know what it should be before trying to restore it.

Auger, or twist bits, are tools that do wear out; they are also easily damaged beyond repair. This is the reason why craftsmen are reluctant to lend twist bits. Added to this, good twist bits have always been expensive. The arrival of powered drilling machines has affected the design of wood- boring

Fig 12.1 Chain saw file (centre) with four engineers' stone files.

tools. It can be very advantageous to study every boring tool that you come across, as this builds up one's knowledge of the cutting techniques used. The knowledge so gained is not only useful when sharpening, but also helps in selecting the best tool for a particular job.

Files are used for sharpening most bits, and the selection of a suitable file is important. Metal-cutting files are tools that do not seem to concern the woodworker as much as they should. Several files should be kept for sharpening bits. Unless the tool to be sharpened is in very poor condition only fine files will be required. A small saw file that can no longer be used on saws because the corners are blunt, can often be pressed into service. There are round files made for sharpening chainsaws which are parallel and not tapered, as is the normal rat tail file, making them ideal for sharpening some types of bit. But if you require

the ultimate performance, several small slip stones will be needed to put the final finish on the tool. Engineers use small, shaped stones they call files which are ideal for use on bits (*see* Fig 12.1). For a fuller discussion of file technology, refer to my book, *Making and Modifying Woodworking Tools* (GMC Publications, 1992).

Bits are not hardened to the same degree as chisels and plane irons, and the metal is easily filed. The resulting edge is not as keen as that of these other tools, nor will it hold an edge that would be acceptable on them. The twist drill I will deal with separately as most are now made from high-speed steel, and anyway, the sharpening technique is totally different. I will deal with the bits in order of their sharpening difficulty, simplest first.

SHELL BITS

The shell bit has a half-cylindrical body which is sharpened on the outside like a gouge. In fact, it so resembles this tool that it is sometimes known as a 'gouge bit'. One would think that this tool would perform better if it was sharpened on the inside, as the edge would then cut the same radius as the stem of the tool. I was so convinced that this would be better that I spent several hours altering a bit. It made no discernible difference. The outside is sharpened with a single bevel at an angle of 30° and a burr will be thrown up on the inside which is best removed with a slip stone (*see* Fig 12.2).

The cutting edge should be straight and at 90° to the length of the tool.

Fig 12.2 Spoon and shell bits.

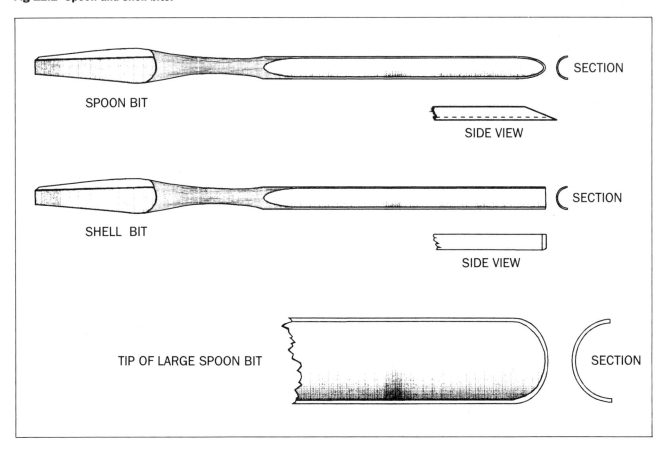

SPOON BIT

SECTION

SIDE VIEW

SHELL BIT

SECTION

SIDE VIEW

TIP OF LARGE SPOON BIT

SECTION

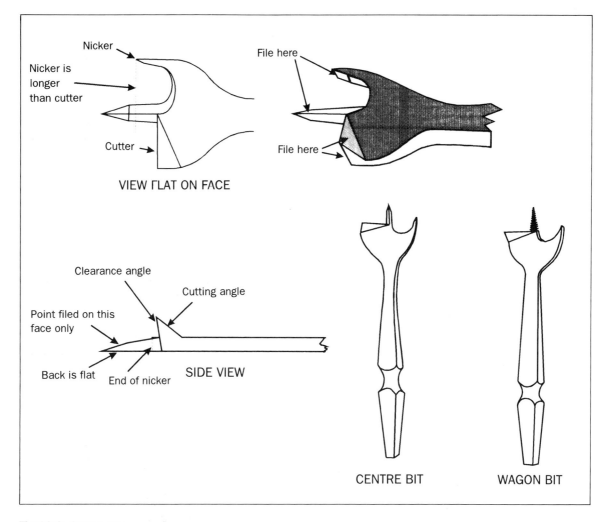

Fig 12.3 Centre bits.

THE SPOON BIT

In years gone by this was a very common tool. Many have survived and can still perform a useful service. The body of the tool is similar to the shell bit (*see* page 95), but the edge is different. This is sharpened to a tapered point, which prevents the tool from wandering when starting a hole. The chair makers used a large version of this bit with the edge sharpened to a more rounded than pointed shape, known as a 'ducksbill'. The spoon bit is sharpened with a single bevel on the outside. The shape of the cutting edge can vary from a pronounced point to a gentle round, depending on the size of the tool: small diameters are pointed and large ones are rounded, and the sizes between graduate from one to the other. Once the outside bevel is filed to shape, the inside burr needs removing with a slip stone (*see* Fig 12.2).

THE CENTRE BIT

We now come to a very useful tool. A centre bit is cheap and very effective for boring clean shallow holes. It is also relatively easy to sharpen. Fig 12.3 shows its various parts. The main thing to be aware of is the cutting circle made by the nicker. This must be outside any other part of the tool, which means that when sharpening the nicker, the outside must be flat; the only bevel here is on the inside. The cutter has a clearance angle on the underside of 2 or 3°, and the upper surface of the cutter is sharpened at 30° to the underside. Point shape is quite important. It has three faces; the one away from the cutting edge is flat and should not be filed at all. The other two faces are the ones that have to be worked upon (*see* Fig 12.4). You may come across a centre bit that, instead of a point, has a screw; this is a very useful item. It is properly called a 'wagon centre bit', and requires less effort in use because the screw pulls the tool into the wood. Unless the screw is badly damaged, do not try to sharpen it. When filing the centre bit use a light hand; the metal is soft and it is very easy to remove too much.

THE HALF-TWIST OR GIMLET BIT

This is a tool which was once made in large numbers. Some survive and may be of use for making screw holes, etc. They have never been very popular as they have a tendency to split the wood, particularly when used near the end of a thin board. They date back to the old button brace and many have a notch filed in the square on their shank where the button latch fitted. A round file is used to sharpen them from the inside. A three-cornered slip stone will remove the burr from the outside, working along the twist to the point (*see* Fig 12.5).

JENNINGS AND IRWIN AUGER BITS

The auger bits most commonly found are of either the Jennings or Irwin patterns. The difference between these two is the spiral that removes the

waste from the hole being bored. Both have two nickers and two cutters, but the Irwin has only one spiral and the Jennings has two. The sharpening technique for these tools is identical. A small flat file with a safe edge is used for most of the work. If you inspect the tool before you start to sharpen it, the way the tool cuts will be apparent. Apart from the screw, it is very similar to the centre bit. The nickers should only be sharpened on the inside; under no circumstances should the outside diameter be reduced. If sharpening the inside of the nickers produces a burr, remove this carefully with a stone.

All bits that have a screw to pull them into the wood are liable to unrepairable damage. Should the screw encounter a nail or other metal object when it is in use, the chances are that the thread

Fig 12.4 Large and small centre bits.

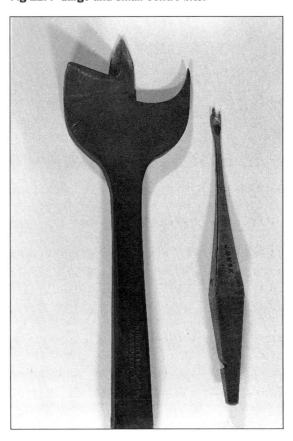

**Fig 12.5
Various bits.**

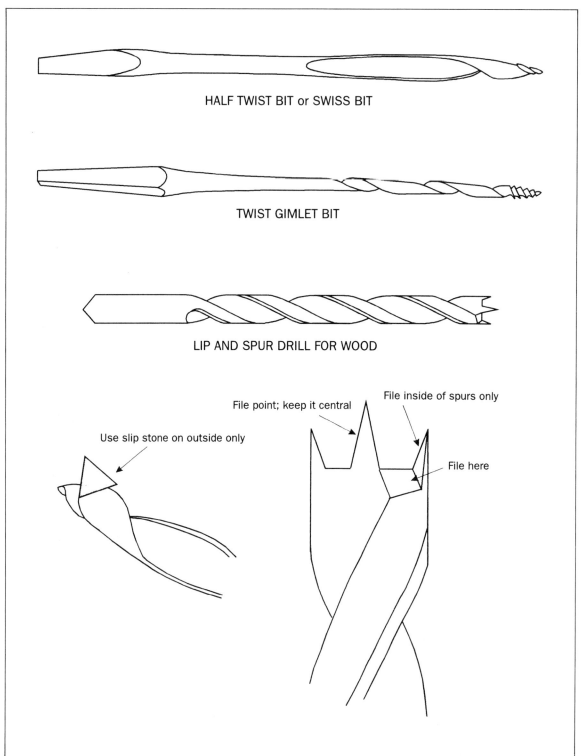

HALF TWIST BIT or SWISS BIT

TWIST GIMLET BIT

LIP AND SPUR DRILL FOR WOOD

Use slip stone on outside only

File point; keep it central

File inside of spurs only

File here

Fig 12.6 Cutting end of a solid-nosed bit.

will be damaged. Careful work with a knife-edged file may put the matter right. There is no metal to spare in the screw and any reshaping reduces its size and its efficiency.

The two cutters need special attention. They are sharpened flat on the underside at an angle to give clearance. The upper face is filed at a cutting angle of 30° to the underside. The edge goes right to the centre where it should intersect with the top thread of the screw. The screw has a double-start thread, one for each cutter. It is difficult to get the two cutters exactly the same length. Particular attention should be paid to this as they should both do the same amount of work. They can be checked by using the tool to bore a hole in a scrap piece of timber. Examine the shaving that each cutter makes. You may find that only one is cutting. It is a matter of trial and error to get them exactly right. Any clean, filed areas of bare metal should have a light coating of oil applied when sharpening is finished, to prevent rusting. Scotch auger bits are sharpened in a similar way to these bits, but having only cutters and no nickers they are somewhat simpler.

EXPANSIVE BITS

There are two separate cutting parts on the expansive bit. The cutting edge on the centre part can be likened to an extended thread at the top of

Fig 12.7 Machine auger bits.

the screw. This is sharpened similarly to the cutters on Jennings or Irwin bits. The detachable blade is filed flat on the underside at a small clearance angle. The top part of the cutting edge is a continuation of a curve and is best shaped with a slip stone. The nicker is sharpened on the inside only. There is no spare metal on any of these parts, so be very careful and remove the minimum necessary. Only sharpen the tool when it is really required.

SOLID-NOSED BITS

The solid-nosed bit is not as common as other auger bits, yet it is unique, because it will bore a hole at any angle to the surface with ease. All other auger bits require a special block or some other device to start an angled hole. A fine-toothed round file is used to shape the cutting edges. There are two cutters and, to ensure that both cut to the same extent, they need similar treatment to that described on pages 97 and 99 for the Jennings and Irwin bits. The outside clearance angle is shaped with a flat file that has a safety edge. The cutting edges should lead into the double-start thread of the screw. Study the tool carefully before starting to sharpen it. The cutting edge is formed around a hole, and a most important point to watch is the way the cutting edge continues around inside the hole. This must cut at the maximum radius of the tool. The clearance angle is only applied at the tip; do not allow it to go up the sides at all (*see* Fig 12.6). Gedge bits are very similar to the solid-nosed bit and their sharpening is almost the same.

MACHINE AUGER BITS

Bits designed to be used in powered machines never have a screw to draw them into the wood. It can be appreciated what would happen if they had. The bit would enter the timber at an alarming rate, and the depth of the hole would be uncontrollable. The machine auger bit is similar to

the Irwin bit, except there is a point instead of a screw, and only one nicker and cutter (*see* Fig 12.7). The sharpening process is almost identical to that employed on the Irwin bit. The only difference is the point: if this is sharpened, it is important that its tip is exactly in the centre, or there will be problems starting the tool accurately. These bits are much cheaper to manufacture than conventional auger bits. They can be obtained with a screw instead of a point for use in a brace.

FORSTNER BITS

The forstner and the saw-toothed machine bit have very similar patterns. The latter is of recent invention and is a derivative of the former. The forstner bit was originally made for use in a brace, but it was found to be very useful in a pillar or pedestal drill. To facilitate the much quicker cutting speed, a small nick was cut in the skirt just behind the cutter. Manufacturers improved on this by extending the nicks completely around the skirt; thus we have the saw-toothed bit. Sharpening these tools is not a task that should be taken lightly. The outside diameter must never be touched, other than to remove any burr formed whilst sharpening the inside. The steel the tools are made from is soft and easily filed. The problem arises when trying to file the bevel that runs around inside the skirt. Movement of the file is restricted by the shallow depth inside the tool, and the file has to be dragged around the bevel rather than using the customary cross stroke. As the tool is mainly used for holes with flat bottoms, the centre point should be kept very small. This tool does not rely on the point to keep it in position; the skirt does this. That is why the bit will cut part of a circle on the edge of a board.

The cutters are filed at 45°, and it is comparatively easy to see that they are both of the same length. If the clearance angle is made about 5° it will be found easy to shape the intersection with the skirt bevel. After much use the nicks around the skirt of the saw-toothed bit may need reshaping with a triangular file (*see* Fig 12.8).

COUNTERSINK BITS

There are two types of countersink used in woodworking: the snailshell and the rose (*see* Fig 12.9). The snailshell is sharpened with a round file inside the cutting edge. The outside shape is not touched other than to remove any burr. On an old tool that has become misshapen it is

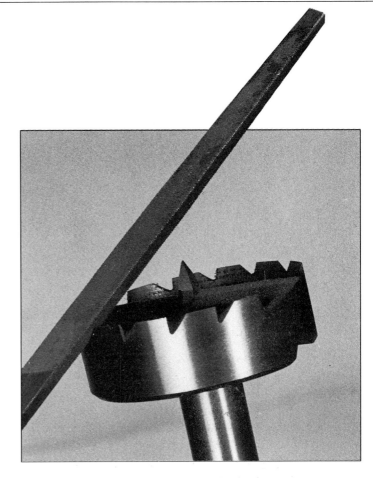

Fig 12.8 Reshaping the nicks around the skirt of a saw-toothed bit with a triangular file.

Fig 12.9 Countersink bits. Left to right: two rose countersinks for use in a brace; two rose countersinks for use in an electric drill; snailshell countersink and flat countersink.

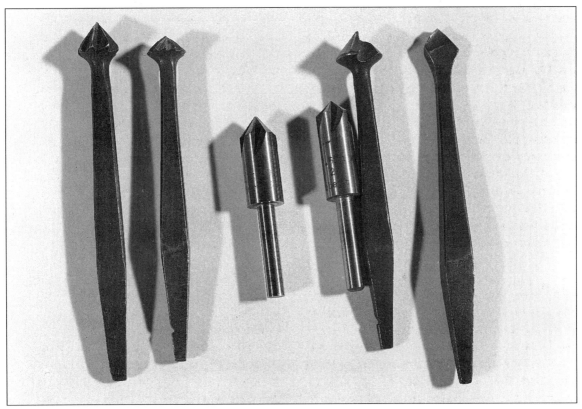

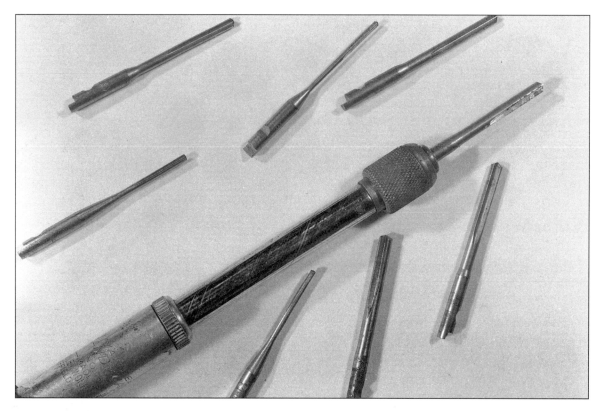

Fig 12.10 Push drill with interchangeable bits.
These are sharpened in a similar way to twist drills.

sometimes possible to grind the outside to the correct cone shape. The cutting edge must stand slightly proud of this cone.

A rose countersink with its many cutting edges may look a formidable item to sharpen, yet it is comparatively easy. Use a small triangular file and give each blade one or two strokes with it. The secret is to file each of the cutting edges the same amount.

THE BRADAWL

A woodworker's bradawl has a sharp cutting edge the full width of its blade. Bradawls with round blades that come to a point are used for leatherwork and have no place in woodworking. The bradawl's chisel-like edge can be sharpened on an oilstone with a flat bevel on each side. In use the bradawl is placed with the blade at right angles to the grain, and pressure is applied while twisting back and forth a few degrees.

THE BIRDCAGE MAKER'S BRADAWL

This tool has a square-sectioned blade that comes to a point. On inspection it may look as though it would easily split the wood; in fact it cuts a clean hole, without any tendency to cause a split. The only sharpening needed is to keep the point sharp. This is best done by rubbing each of the four sides on an oilstone.

PUSH BITS

The small diameter bits made for use in spiral screwdrivers and special push drills (*see* Fig 12.10) need very infrequent attention. The tip of the bit only is sharpened. The cutting point resembles the tip of a twist drill (see below), and it should be sharpened in exactly the same way.

TWIST DRILLS

There are two distinct patterns of twist drill. The tip design varies depending on what material the

Fig 12.11 Twist drill angles.

Fig 12.12 Martek twist drill sharpening attachment for electric drill.

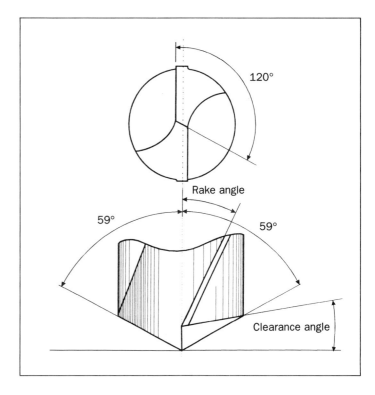

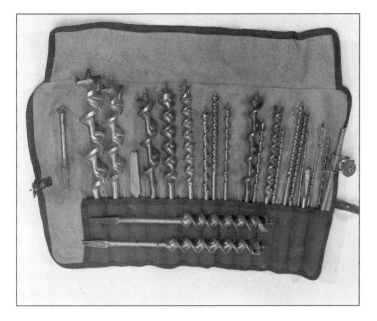

Fig 12.13 Twist bits in a bit roll; a good form of storage.

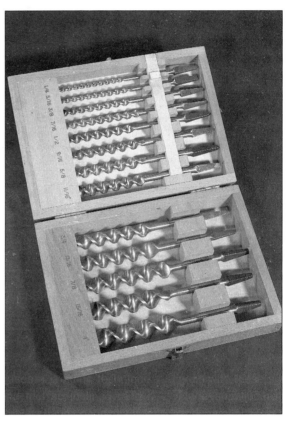

Fig 12.14 A set of twist bits in a fitted case.

drill is intended to work in. The common high-speed twist drill, as sold in tool shops, has a tip that is designed to drill holes in metal, and the other design, made for drilling holes in wood, has a very pronounced point. The former will work perfectly well in wood as well as metal, but the latter will only work in wood. The point on the woodcutting twist drill stops the drill wandering when it is started, and makes it easier to position the drill accurately on a marked spot when dowelling. As both these drills are usually made from high-carbon or high-speed steel, they are sharpened on a grindstone. The corner of the wheel is used to sharpen the point of the wood drill. The original bevels and shape must be carefully maintained (*see* Fig 12.5).

Fig 12.11 shows the bevel angles at the tip of a twist drill. It is possible with practice to grind these by hand, although most people find that some form of jig or attachment is necessary. Fig 12.12 shows one such attachment.

CARE OF THE TOOLS

Bits are best kept in a bit roll (see Fig 12.13). A complete set of bits are sometimes housed in a fitted case (see Fig 12.14). This is fine when they are only used in the workshop, but for tools that need transporting, the bit roll is a much better idea. Forstner bits are normally kept standing in a block of wood. Their shanks are nearly always the same size regardless of the size of hole they cut. It is an easy task to bore a series of holes in a block of wood to make a stand for them. Twist drills can be bought as a set in a metal case. This is ideal if they are stored in a drawer, or need to be transported about. However, the drill stand is a much better proposition in the workshop.

CHAPTER 13

SHARPENING MACHINE SAW BLADES

SAFETY

When working with woodcutting machinery one must be aware of the inherent danger of all machine tools. Safety is mainly a matter of common sense, yet there are several things of which one needs to be aware. While sharp hand tools may inflict a nasty cut, a machine can amputate a limb. When removing or installing blades, the machine should be electrically isolated. Fixed machines should be fitted with an isolator, and portable machines should be unplugged. This advice may just seem common sense, but after nearly 50 years in the trade I have come to realize that most accidents are caused because people do not always use their common sense.

Blunt teeth on a saw will rapidly build up heat caused by friction. Heat will build up around the periphery of the saw. The centre has flanges and a nut fixing it to the arbour; any heat at the centre is lost in all this metal. Therefore the hot rim of the blade expands while the centre remains stable. The only way the blade can accommodate this expansion is for the edge to take on a wave form. The saw will now be unable to cut straight, and will probably bind in the kerf. The wood is then gripped tight by the blade and thrown up at the back of the saw, pulling the operator's hands onto the blade. Sometimes the rim becomes so distorted that the saw does not return to its original shape on cooling. You will see from this that not only is the quality of work affected by a

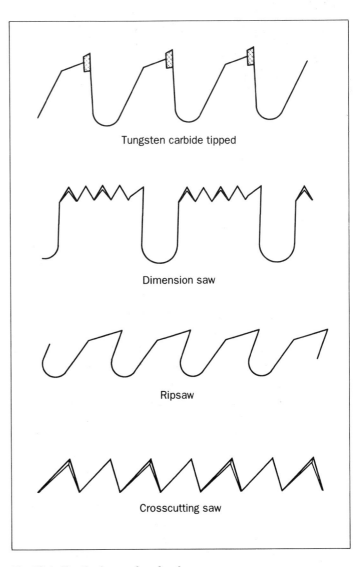

Fig 13.1 Tooth shapes for circular saws.

blunt saw but it is also dangerous.

Looking at a circular saw blade one is inclined to think it a very simple tool: just a flat plate with teeth around its periphery. It is in fact called a saw plate in the trade, and the technology behind this simple piece of apparatus would fill a book. The size of machine found in most small workshops does not require the operator to be conversant with all this information; much of it only applies to the large diameter saws. However, if the machine is to perform at its best and produce clean-cut, accurate work, some understanding of it is necessary.

To start with, the machine needs to be set up correctly. For instance, is the fence parallel to the blade? Is the blade parallel to the guide grooves in the surface of the bench? Is the riving knife of the correct thickness for the saw kerf, and set in the right position? If the saw is of sufficient size to need packing, is this in good condition and correctly installed? Was the saw mounted on the arbour correctly? Is the speed of the saw correct? All these things affect the efficiency of the saw, and even a superbly sharpened saw cannot operate properly if the machine is incorrectly set up. There are several good books that cover these features (see page 147).

CIRCULAR SAW BLADES

Circular saw blades have teeth of different shapes depending on what they are to be used for, because no particular tooth shape will cut at its best on all types of work. Saws in the sizes being described here fall into five distinct groups. These are ripsawing, crosscutting, dimension sawing, delicate cutting (plywood, etc.), and abrasive material cutting. Fig 13.1 shows the different configuration of four of these types. In Chapter Eleven, when discussing hand saws, I described the difference between ripping and crosscutting with a hand saw (see page 82). The circular saw behaves in a similar fashion.

Ripsaw teeth are sharpened with a chisel edge, and crosscut teeth with a point, almost the same as a hand saw. The rake of the teeth is also similar; rip teeth have a positive rake and crosscut teeth a negative. Added to these two basic types we have the cut-off saw, used where the work is presented to the saw at an angle. This saw can cut at any angle to the fibres of the wood. Some people refer to the cut-off saw as a 'multi-purpose saw blade', which it is not. It is also sometimes called a 'novelty saw'.

When cutting plywood or laminates, fine teeth are required to prevent the lower surface from breaking out or becoming whiskered. The teeth on these saws are formed with a neutral rake.

Tungsten carbide-tipped (TCT) saw blades have the cutting face of each tooth formed from tungsten carbide. This is a very durable material with an extremely high wear resistance. A TCT saw blade remains in cutting condition far longer than a standard steel blade. Where abrasive timbers such as teak are used, the TCT blade soon recoups its additional cost.

The blade size is classified by the diameter of the plate. Large diameter saws need special treatment. Fortunately this large size of saw is rarely found in the small workshop. The following descriptions apply to saw plates of 18in (457mm) diameter and smaller. Anything over this size should be entrusted to a professional saw doctor. These larger plates need pre-stressing, known as **tensioning**. This is done by placing them flat on a crowned anvil and striking with a dog-head hammer. Not a task to be undertaken unless one knows exactly what is required. Tensioning of saws is perhaps the most difficult of all tool maintenance and calls for a high degree of skill and experience.

Three things must be attended to if the saw is to be kept in good working condition: the teeth need to be kept in a true cutting circle; the set that gives the saw its side clearance must be maintained; and thirdly, the teeth must be sharp and of the correct shape. I will deal with each of these requirements separately.

STONING

To keep the saw round a treatment known as **stoning** is used. This entails using an abrasive stone to touch the teeth lightly while the saw is running. It sounds like a dangerous and hair-raising thing to do, but if it is done properly there is no danger. Do not use a flat bench stone as is sometimes done; it is difficult to hold, and the hand is too near the saw. Mount an old grinding wheel on an angled wooden block at the end of a board (*see* Fig 13.2) using a bolt and washer. As the wheel wears it can be turned around so that a new face is presented to the saw. The board should be wider than the diameter of the wheel so that, when the board is placed flat on the saw bench, the edge can be against the fence. It is then advanced gently until the stone lightly touches the revolving teeth. The stone is withdrawn and a short cut made in the end of a piece of scrap wood to see if the end of the cut is square. A square cut shows the teeth on both sides of the saw have been stoned an equal amount. If the end of the cut is out of square the fence may be adjusted so that the stone is repositioned. Stoning should put a small flat on the tip of each tooth. This flat acts as a datum

Fig 13.2 Old grinding wheel used for stoning circular saw. Riving knife and crown guard should be attached to machine while stoning. They have been omitted from this drawing for clarity.

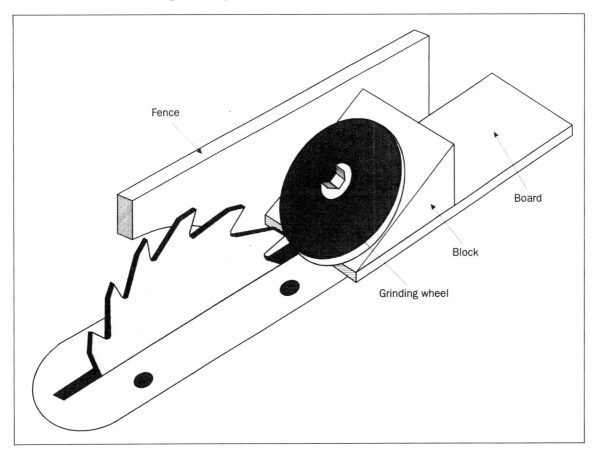

Fence

Board

Block

Grinding wheel

when sharpening. Before removing the blade from the machine, inspect it to see that every tooth has a small flat on it.

There is an important requirement overlooked by many sawyers: the hole at the centre of the saw blade must be a clearance fit on the arbour. This clearance will vary from blade to blade; as well as this, the arbour may be slightly undersize. When a blade is mounted it should always be at the same orientation or it will not necessarily run concentric with the arbour. Machines with a drive pin on the fixed flange, that locates in the drive hole of the saw blade, should have the pin at the top when mounting the saw. Unfortunately most saws under 12in (305mm) in diameter do not have a drive pin. Mount these saws with the maker's name at the top, or put a small centre punch mark near the hole. Only a small mark is needed; be careful not to distort the saw plate. If another centre punch mark is put on the rim of the fixed flange, the blade can be accurately located at the same orientation to the arbour every time you mount it. The centre pop mark on the flange is positioned at top dead centre when mounting a saw. The centre pop mark on the blade is positioned against that on the flange. This will ensure that the weight of the blade positions the saw exactly the same every time you mount it.

Before stoning the saw for the first time, be sure the saw is mounted as described above. It requires only a very small discrepancy in the concentricity of the teeth for most of the work to be done by only half the teeth. The correct setting up of the blade in this way makes for a much sweeter cutting machine. The effect is noticeable in the surface of the cut material.

SHARPENING

Saws can be sharpened by a machine that uses a special-shaped grinding wheel, but a much better edge is obtained by hand filing. The file used for most saws is called a mill file – a single-cut flat file with half-round edges. There is a special piece of apparatus known as a **saw horse** for holding the plate while it is being filed. However, in most workshops, two boards with their ends rounded are placed either side of the plate. A bolt passes through the hole at the centre of the saw to clamp them together. A strip of hard rubber or some similar resilient material is glued to the inside face of the boards at the rounded end. The boards with the saw clamped between them are held in the vice at a suitable working height.

Whatever the tooth shape might be, the flat produced by stoning is removed when filing, ensuring that the teeth cut in a true circle. Keeping the gullet at an even depth right round the plate is not easy. About every fourth or fifth sharpening I coat one side of the plate around the edge at gullet depth with engineers' setting out blue. A line with a pair of dividers can be marked into the blue at a new gullet depth. It is better to mark the gullet depth before removing the plate from the machine. A scriber is held flat on the saw table with the point against the blade and the saw revolved by hand. This puts a line on the saw true to the cutting edge. With the round edge of a mill file the gullets are reshaped to the new depth. I have a narrow, round-edged grindstone that is very useful for gulleting saws, provided they have large teeth. If you try to do the same be very careful. Make sure the tool rest on the grinder is set to grind the blade at 90°, and practise presenting the saw to the wheel before turning the power on. The function of the gullet is to clear the sawdust from the cut. Should the gullet be reduced in size through constant sharpening of the top of the teeth only, the saw's efficiency will be drastically reduced.

Saws used for fine cutting, and the triangular teeth of the cut-off saw, are sharpened with a three-cornered saw file. The technique is almost identical to sharpening a hand saw (see page 92). With the blade mounted between the boards as described above, secure the assembly in a vice. The best height for filing is a little above elbow height. This enables the tooth to be seen clearly

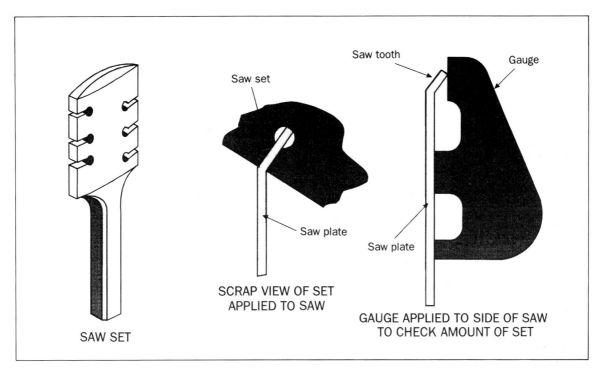

Saw tooth

Gauge

Saw set

Saw plate

Saw plate

SCRAP VIEW OF SET
APPLIED TO SAW

GAUGE APPLIED TO SIDE OF SAW
TO CHECK AMOUNT OF SET

SAW SET

Fig 13.3 Saw set and gauge for circular saw.

with no difficulty in maintaining the correct file angle. Fig 13.1 shows all the relevant tooth angles. Saws used for cutting softwood can be sharpened with a larger angle than those used for cutting hardwood. The raker tooth on the cut-off saw is kept below the cutting circle of the other teeth by about ¹⁄₆₄in (0.4mm). Just give it an extra two strokes of the file after the flat is removed.

Blades used for wobble sawing, and those in dado sets, are best inspected for tooth configuration before sharpening. There are a number of variations and the original tooth configuration should be maintained. As a saw blade is repeatedly sharpened its diameter is reduced. The periphery speed of the saw is dependent on the diameter. Eventually the size will be reduced to the point where the speed is below the optimum, which is about 10,000ft per minute (3,048m per minute). At some point the saw's performance will indicate that the time has come for it to be replaced. Periphery speed is not

Fig 13.4 Device for setting the teeth on a circular saw.

Fig 13.5 Saw plate being set on appliance pictured in Fig 13.4.

only related to the speed at which the wood is cut. It greatly affects the quality of the cut, and the safe working of the machine. Machines used where the Woodcutting Act applies must have the minimum size of blade displayed on the machine.

It is impossible to file the teeth of a TCT saw blade. These have to be ground with a silicon carbide wheel, and it is a job best entrusted to a company that has the necessary equipment to perform this task. They are also able to replace a damaged tooth face by brazing a new tungsten tip on the plate.

SETTING

The width of saw kerf to allow for blade clearance must be maintained. There are three methods that

Fig 13.6 Swage setting on saw blade

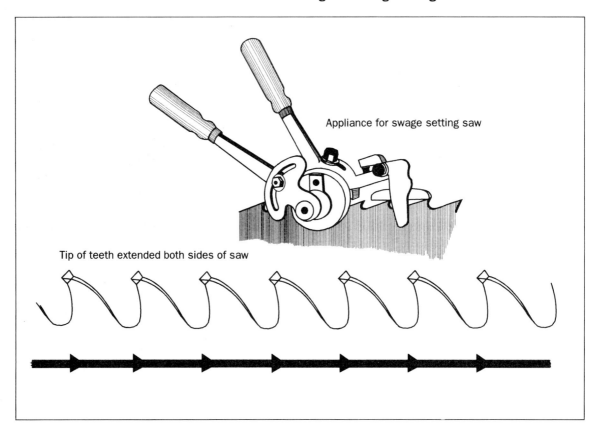

Appliance for swage setting saw

Tip of teeth extended both sides of saw

are used to apply set to the saw. There is a fourth that involves using a nail punch on each alternate tooth with the saw laid flat on a piece of hardwood, but this is not to be recommended as there is no control over how much set each tooth gets. The first method involves a tool (*see* Fig 13.3) which is used with a gauge to bend the tip of the tooth the required amount. The use of this tool and gauge requires some skill in the judgement of how much to bend each tooth. Over-bending and consequential correction can break a tooth out of the saw. A second and much better method is to use the machine shown in Figs 13.4 and 13.5. This is adjustable and will handle most saw diameters up to 18in (457mm). The amount of set is also adjustable, depending on how the anvil is set. Thirdly, there is swage setting, which requires a special tool that shapes the tip of each tooth. Instead of bending one tooth to the left and the next to the right and so on, this tool applies set to both sides of the same tooth (*see* Fig 13.6).

BAND SAW BLADES

Band saws can be divided into two separate classes: narrow band saw machines that take blades from ¼in (6mm) to 1½in (38mm), and the wide band saws with blades from 3in (76mm) to 12in (305mm). The latter, which are used for conversion and resawing, are always entrusted to a saw doctor, and the blades under ¾in (19mm) are so priced that sharpening is uneconomic. Most narrow band saw blades have hardened teeth that cannot be reset or sharpened. Blade is best bought by the 100ft (30.5m) coil and made up in the workshop, an electric brazing machine (*see* Fig 13.7) being used to join the blade. It is possible to braze the blade using a gas torch and a jig. I am not in favour of naked flames in my workshop, however, I know of several woodworkers who use this technique. Where band saw blades are only occasionally joined, the cost of a machine may be out of the

question, in which case a simple jig can be made from a short length of angle iron. What is needed is a means of holding the spliced ends together and keeping the blade straight while the brazing is carried out. It is probably wrong to call this joining technique brazing as it is actually done with hard solder. If attempting this method, do not cool the red-hot blade with water, or you will make it hard and brittle. Many joints fracture because the blade is brittle and it is unable to flex as it passes around the wheels in the machine. Another method of joining the blade can be done with a butt-welding machine, which is often part of the band saw machine. The blade joints and methods of making them are shown in Fig 13.8.

It is possible to hand sharpen some narrow band saw blades. Sharpening is usually carried

Fig 13.7 Machine for brazing narrow band saws.

Fig 13.8 Brazed band saw joint.

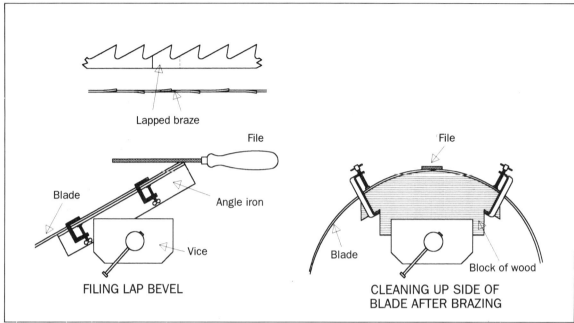

Lapped braze

File

File

Blade

Angle iron

Vice

FILING LAP BEVEL

Blade

Block of wood

CLEANING UP SIDE OF
BLADE AFTER BRAZING

out on a machine that uses a grinding technique. If you have the time and feel so inclined, and if the teeth are of a suitable shape, hand filing is possible. Fig 13.9 shows all the necessary angles. The techniques employed are very similar to those used to sharpen hand saws.

Holding the blade while sharpening can be a problem. A saw block (as used for hand saws) against the bench works quite well. The loop of blade not in the block can then lay safely on top of the bench. Setting can be carried out using a pliers saw set. There is, however, a special setting tool made for band saw blades (*see* Fig 13.10).

CHAIN SAW BLADES

Whether the chain saw can be classed as a woodworking tool is debatable. However, some people have produced carvings with one. I use one in my timber store to crosscut large mahogany boards. There are several systems made that use a chain saw for converting timber from the round. Different blades are used for ripping down the

grain with these systems than the normal crosscut blade supplied with the machine. Special round and parallel files are made for sharpening chain saws, available in different sizes to match the tooth sizes of the different blades. The cutting teeth of the chain saw have a different cutting action to other saws. Each tooth – usually referred to as a cutter – is similar to a small gouge. It scoops chips from the wood being cut. This is evident when you look at the sawdust from a chain saw which is more like small chips than dust.

There is a sharpening jig made by Omark which simplifies the sharpening process (*see* Fig 13.11). Unless you have received training in the use and sharpening of the chain saw, I would say this appliance is a necessity. When filing, it is very difficult to maintain the cutting edges of the teeth at the complex angles required without some form of mechanical assistance. One attraction of sharpening the chain is that it is not removed from the machine. Should you attempt the sharpening, make sure that all the original angles are kept. There are two different shapes of tooth: square-

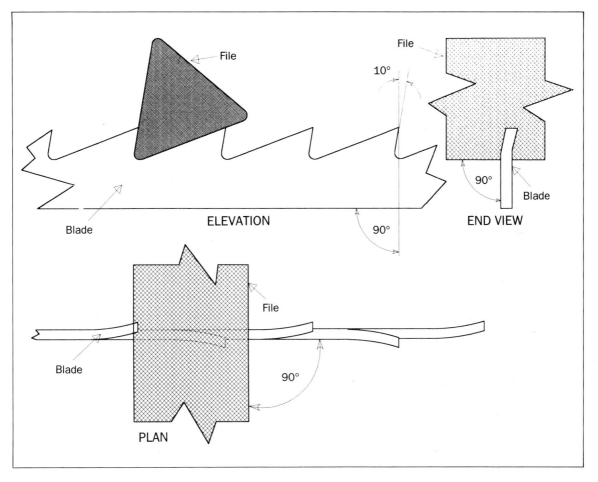

Fig 13.9 Sharpening a band saw blade with a file.

topped cutters and round-topped cutters. The square top is more like a chisel while the round top is like a gouge. There are different sizes of cutter that require different sizes of file (*see* Fig 13.12).

CARE OF BLADES

Circular saw blades are best hung up on the wall when not installed in the machine. Keep the teeth clean; resin that builds up on their sides when working softwood is easily removed with a stiff brush and some paraffin (kerosene). Narrow band saw blades should be coiled up using a technique

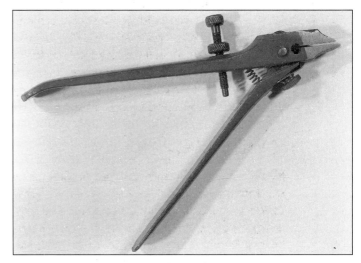

Fig 13.10 Saw set for narrow band saw blades.

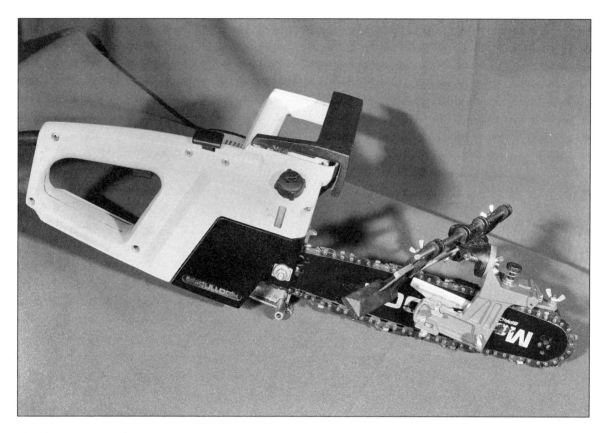

Fig 13.11 Chain saw sharpening jig fitted on saw.

known in the trade as 'folding', where the blade is coiled into three loops. Once you have mastered the technique it is very easy, but the first few times it seems like a Chinese puzzle. Hold the saw in two hands with about two-thirds of the saw below, and the teeth pointing away from you. Use the fingers and thumbs to twist the teeth outwards. Persuade the loop formed at the top outwards, and down, so that it meets the bottom loop, with the teeth facing one another. The hands are brought together and the two half loops passed over one another. The bottom band forms two loops which lay on the floor. The top loop is brought down on top of them. The blade is now in three loops with all the teeth facing the same way. (*see* Figs 13.13 to 13.21). They can then easily be hung up out of harm's way.

File sizes for chain saws

Chain Types: 25AP, 33LG, 35LG, 91G, 91SG.
Use ⁵⁄₃₂in (4mm) file.

Chain Types: 19AP. 20AP, 21AP, 76LG, 78LG.
Use ³⁄₁₆in (4.8mm) file.

Chain Types: 26(P), 27(P), 28(P), 50C(P), 51AC(CP), 52AC(CP), 50L(LP), 521(LP), 72D(DP), 73D(DP), 75D(DP), 72LP, 73LP, 75LP.
Use ⁷⁄₃₂in (5.5mm) file.

Chain Types: 9AC, 10AC.
Use ¼in (6.3mm) file.

Chain Types: 11BC.
Use ⅜in (9.5mm) file.

Fig 13.12 File sizes for chain saws.

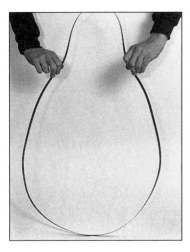

Fig 13.13 Take the band saw and hold as shown with your thumbs towards you.

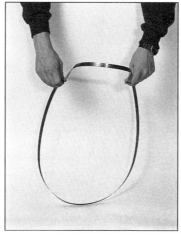

Fig 13.14 Twist your thumbs towards one another, forcing the top loop of the blade away from your body.

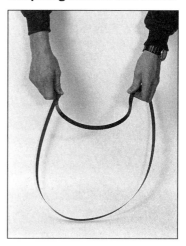

Fig 13.15 Continue forcing the top loop downwards.

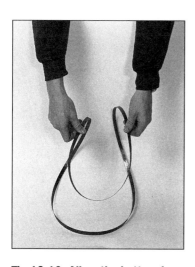

Fig 13.16 Allow the bottom loop to lay parallel with the floor.

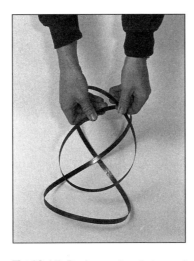

Fig 13.17 Push your hands towards the floor and the bottom loop.

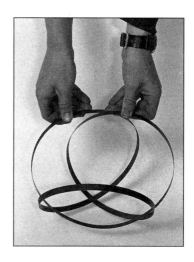

Fig 13.18 Tip the top loop into the bottom loop.

Fig 13.19 Bring the hands together.

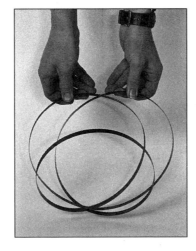

Fig 13.20 Keep the loops pushed down on the floor.

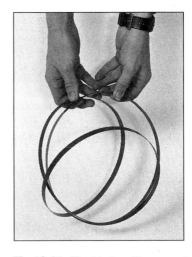

Fig 13.21 The blade will now spring into three loops, making it much easier to store.

CHAPTER 14

SHARPENING MACHINE TOOL CUTTERS

KEEPING IN BALANCE

The balance of cutters that revolve at high speed is very important. Most people have driven a car with a tyre that is out of balance, and have experienced the vibration caused to the steering. The car wheel revolves at a slower speed than a woodcutting machine's block. A small imbalance may not be noticeable but will nevertheless load the machine bearings in a way they were not designed for. All forces acting on a block in motion must be balanced by equal and opposite forces. If all the cutters in a block are of the same size and weight, and set to project the same amount, the block is in **static balance**. This does not necessarily mean that it will be in balance when it is running at high speed (**dynamic balance**). Once the cutter block starts to rotate, **centripetal** force is present, stopping the cutter flying outwards. It is a basic law of motion that, once in movement, a body will continue to move in a straight line and will resist efforts to change its path. The cutter is restrained from flying off and continuing in a straight line by the nut or bolt that secures it to the block. This restraint, exerted by the nut or bolt, is the force continually required to bend the cutter from its straight path. The greater the speed of the machine the greater the desire of the cutter to fly off in a straight line. The force which acts to move away from the centre is **centrifugal** and must be balanced by an equal and opposite (i.e. centripetal) force.

The quoted speed in rpm for the cutters in a block is inversely proportional to the cutting diameter. A 6in (152mm) Whitehill disc type block would ideally run at 6,000rpm. A little simple arithmetic will show that the cutters in this block are travelling at 126mph (203kph). Out of balance measured in milligrams at standstill can be measured in kilograms at this sort of speed. Out-of-balance cutters will inevitably affect the quality of the work being produced, and will also place undue strain on the spindle and bearings. Cutters that are severely out of balance may be thrown from the block, and fly across the workshop like a projectile.

Multiple cutter blocks need to be set so that all the cutters do the same amount of work. This even applies to solid router cutters. The profile of each blade must cut exactly the same path as any other blade in the tool. This is not always easy to achieve without sophisticated equipment.

Blunt cutters will rapidly build up heat caused by friction. Keep all machine cutters sharp, and feed at the correct speed as they rely on entering uncut timber to keep cool. Feeding too slowly is as dangerous as feeding too fast.

PLANING MACHINE CUTTERS

Most planing machine blocks have two or three cutters. These must be kept in sets. If the cutters are kept in sets and all the cutters in a set are treated the same they will remain in balance. The

Fig 14.1 The Sharpenset whetstone grinder, fitted with the planer blade grinding attachment made by myself. The attachment supplied by the manufacturer left so much to be desired, I made this one from solid phospher bronze.

Fig 14.2 Home-made jig for grinding planer blades on a pillar drill.

cutters removed from a machine when changing to another set should be put in a specially made case that will keep them together and protect their edge. If you have some accurate means of weighing them, static balance can be checked (all the cutters in the set weigh the same). When grinding the cutters, care is needed to remove the same amount of metal from each cutter in the set.

Cutters – sometimes called knives – can be made from one of the following: high-carbon steel, high-speed steel, tungsten carbide tipped. Each requires a different type of grinding wheel (*see* Chapter Three). There is no way that planer cutters can be ground freehand, and many grinding machines have attachments specially designed for holding them in place. Some of these devices leave a lot to be desired. I made

my own (*see* Fig 14.1) because I was so unimpressed with what was on offer by the machine manufacturer. I have also made a device for grinding cutters on the pillar drill (*see* Fig 14.2). Some machines have disposable cutters; no attempt should be made to sharpen these.

Most cutters have a single bevel on one side, usually 30°, and the other side remains flat. Grinding follows the principles stated in Chapter

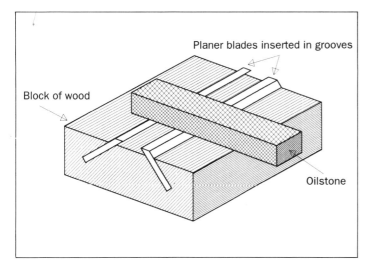

Fig 14.3 Method of honing planer blades.

Three. The three things that must be maintained are the bevel, straightness of the cutting edge, and the balance. The ground cutters should be honed. There is no separate honing angle on machine cutters; two cutters are put in a block as shown in Fig 14.3 and the stone rubbed flat across both simultaneously. A diamond-plated sharpening stone is the ideal tool for this operation; if you do not have one, an ordinary oilstone will do just as well, but it will take longer to get the required sharpness.

Cutters are often honed while they are in the block to restore a sharp edge. There is a device made by DMT (*see* page 146) for honing planer cutters on their diamond sharpening stones. Diamond sharpening systems have much to

Fig 14.4 Spindle moulder cutter blocks.

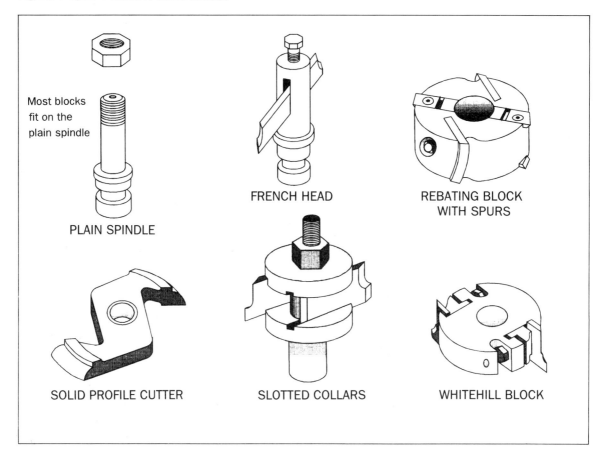

recommend them when it comes to maintaining sharp edges on machine tools (*see* Chapter Five).

THE SPINDLE MOULDER

Cutters used in the spindle moulder come in many shapes, sizes and designs, and it is impossible to give details of all of them here. The main types of cutter are the French head, square block, Whitehill, rebating block with spurs, slotted collars, solid and throwaway tipped (*see* Fig 14.4). The last two cannot be sharpened in the small workshop. The throwaway tipped, as its name suggests, has disposable tips. Solid profile cutters need special machinery to grind their profile. They are seldom found in the small workshop; their main use is where many thousands of feet of the same profile mould are continuously needed.

Cutters for use in the French head are best made double-ended if possible, to keep the cutters in balance. Single-ended cutters must have blanks inserted to balance them. The advantage of the French head is that the cutter's profile is the reverse of the moulding to be cut. This is not so with any of the cutter blocks, and some simple geometry is needed to determine their correct shape (*see* Fig 14.5). Cutters of

Fig 14.5 Obtaining the profile of a cutter to cut a given section of moulding.

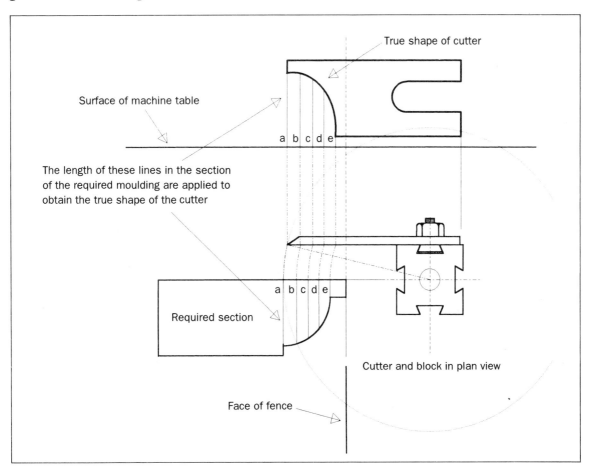

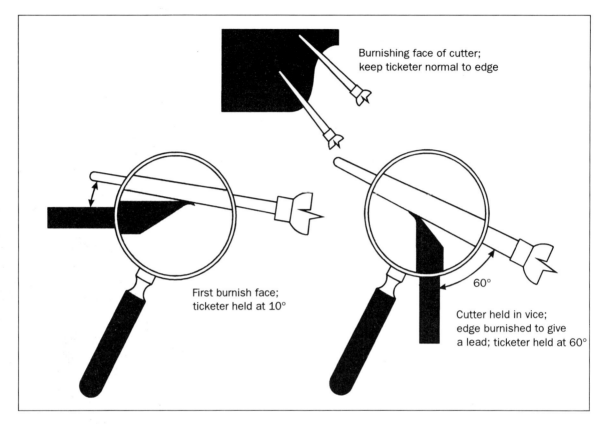

Burnishing face of cutter; keep ticketer normal to edge

First burnish face; ticketer held at 10°

60°

Cutter held in vice; edge burnished to give a lead; ticketer held at 60°

Fig 14.6 Burnishing French head cutter.

complex profile are ground to shape on a bench grinder with narrow wheels that have a rounded edge. Here again, balance is important; keep all the cutters in a block exactly the same. The cutter is ground and honed at the same angle, usually 30°. Rub the face of the cutter flat on an oilstone to remove any burr. The edge of French head cutters needs some special treatment because the cutters are presented at a right angle to the stock being worked. The cutters are not hardened. They can easily be shaped with a file and oilstone slip. When they have been brought to the correct profile with a sharp edge, they are burnished (*see* Fig 14.6). This gives some lead to the cutting edge.

Setting blocks up so that the cutters project the same amount is simplified if the block is supplied with a setting device (*see* Fig 14.7).

Some workshops use a dummy spindle to mount the block on while adjusting the cutters. This consists of a spindle the same diameter as that in the machine, mounted on a metal base. An adjustable pointer is set up and the cutters adjusted so each just touches it.

HOLLOW MORTISE CHISELS

The hollow mortise chisel can be sharpened by filing the inside and then rubbing the four outside faces flat on an oilstone. Using this process, it is difficult to maintain the shape of the cutting edge. A special tool is made which resembles a countersink. This tool has a locating pin on its nose which fits into a bush that is placed inside the chisel. The tool is kept exactly central by the bush and pin. One tool sharpens all sizes of chisel, but a separate bush is needed for each size.

ROUTER CUTTERS

Router cutters are classified by the material from which they are made: high-speed steel, tungsten carbide tipped, solid tungsten, or diamond-bonded. Solid tungsten cutters are expensive, diamond-bonded even more so. The latter are only worth buying if the tool is used for high volume production. High-speed steel cutters are the cheapest, but have a short life. As time goes by there are less of these in my workshop as I find TCT cutters outlast and out-perform them. The additional cost of TCT is repaid many times over in the amount of downtime saved.

There is a difference between sharpening and grinding cutters. Sharpening with a slip stone is carried out by rubbing the flat inside face of the cutter only. The outside profile and clearance angle are not touched. Precision equipment is required to grind cutters if they are to be maintained in first class condition. The cost of the machine for carrying out this task precludes its installation where it will not be in continuous use. Most saw doctors offer a cutter grinding service. I cannot overemphasize the merit of having cutters maintained by a specialist firm. The small cost of the service compared to the cost of a new TCT cutter makes it particularly attractive.

There are attachments available to fit some hand routers for grinding straight-edged cutters. The time and trouble setting up to get an accurate ground tool is considerable. High-speed steel cutters need frequent attention if they are to be kept in a reasonable working condition. They can be sharpened with an oilstone or diamond slip, and when doing so it is most important to keep all the angles correct. It is best if only the inside flat face is worked upon, so that clearance and cutting angles are not upset.

As a cutter is repeatedly ground and sharpened, its cutting circle is reduced (*see* Fig

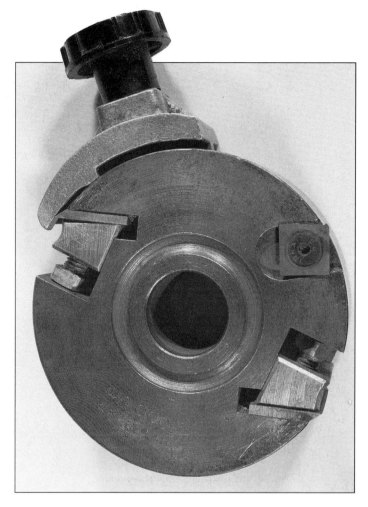

Fig 14.7 Cutter setting device attached to rebating block.

14.8). This must be remembered when using grooving cutters, particularly those used for grooves to fit ply into, for instance in a drawer side. We all have standard cutters for this type of task. After they have been sharpened or ground, run a groove in an offcut, and check that the ply still fits into it.

SLOT MORTISING CUTTERS

There is some confusion over the name of this tool. The cutter for what is properly called the 'oscillating bit mortiser', is known to some as the slotter or jigger. It has two distinctly different

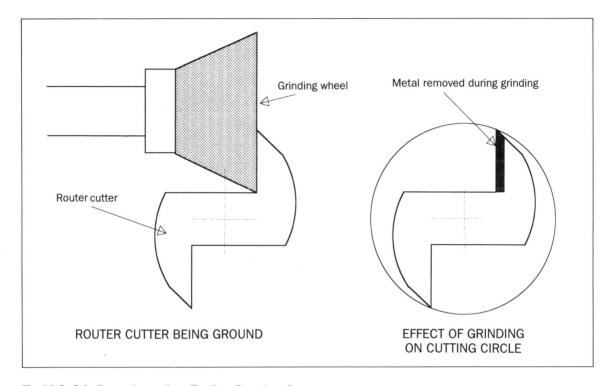

ROUTER CUTTER BEING GROUND

EFFECT OF GRINDING
ON CUTTING CIRCLE

Fig 14.8 Grinding reduces the effective diameter of router cutters.

Fig 14.9 Slot mortise cutters.

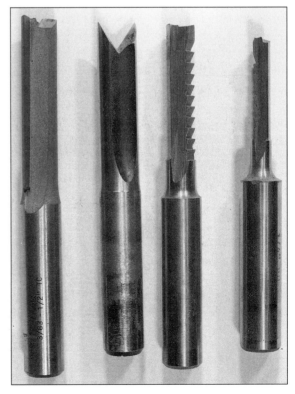

designs (*see* Fig 14.9). On the original design, there is no way of mechanically removing the chips from the mortise being cut, and this is the reason for the teeth on the improved design. The teeth masticate the chips and the resultant smaller particles easily escape from the mortise as it is cut.

The standard cutter is only sharpened on the end. There are four ground faces, which can easily be seen if you inspect the tool (*see* Fig 14.10). Most of the work is done by the two points formed by these ground faces. Sharpening these faces with a slip stone is all that is required. Try to keep both points the same length.

The improved pattern has two ground angles at the tip. These can be reground if care is exercised. The flat inside face of the plain cutter and that of the teeth are sharpened with a slip stone, as one would sharpen a router cutter. The outside face of either cutter should not be touched except to clean any resin or other deposit from it.

CUTTER CARE

Cutters, when they are not installed in a machine, need special treatment. It is so easy to throw them into the bottom of a drawer. Spindle moulder cutters should have some form of storage that keeps them separated from one another. No, they don't breed if they get together, unfortunately. What does happen is the edges become knocked and gapped. It takes only an hour or so to make a fitted box (*see* Fig 14.11). Cutter blocks can be hung on the wall. If they are not used for several months they should be inspected for rust. Any signs of corrosion should be removed and a thin film of oil applied.

The knives from planing machines should be bolted to a board. This works fine with a two knife block, but when one has a three knife block the

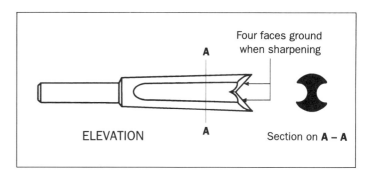

Fig 14.10 Slot mortising bit.

third knife has to be given a board of its own. Otherwise it can very easily be separated from the rest of the set. To ensure my knives stay in sets I stamp each blade in the set with the same number, and these are always checked when installing or grinding. Resin and other deposits should be cleaned from all cutters before they are sharpened or ground. Blades installed in a machine should also be kept clean. Most deposits can very easily be removed with an old paint brush and a copious supply of paraffin.

Fig 14.11 Fitted box for storage of spindle moulder cutters.

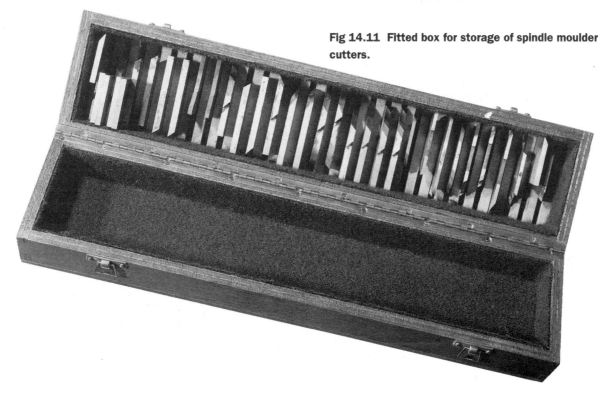

CHAPTER 15

Sharpening Scrapers, Knives and Other Tools

The Cabinet Scraper

How is it that the simplest of all the woodworking tools is probably the hardest to sharpen? I have met a number of people who, having obtained a cabinet scraper, have given up trying to use it. It is such a useful tool that I urge every woodworker to persevere and master the sharpening technique,

which is very simple. The problem is one of applying just the right amount of pressure when burnishing the edge.

As with all tools it is important to know what the correct working edge should look like. The scraper has a different cutting action to most other tools. The edge is a hook (*see* Fig 15.1) and one would assume that the bigger this hook the better the tool would work. This is not so, however, for there is an optimum size. The best way to find this optimum is to experiment. Scrapers used in planes and other tools have a much larger hook than the hand-held cabinet scraper, and they are sharpened differently. First I will deal with the cabinet scraper, which is just a piece of flat steel. The most usual size starts at 6in long by 2½in wide (152mm by 64mm) and 22 gauge thick. I say 'starts at' because as the scraper is constantly being sharpened, the width is reduced.

The first step in sharpening the cabinet scraper is to file the edge. This is best done by holding the scraper in a vice. A medium-sized mill file is used to make the edge straight and square. The file leaves marks that need to be removed by rubbing the edge on an oilstone. Take care to keep the edge square. As this process can wear a groove in the stone it is best to use the side. If the oilstone box lid is left on the stone, the scraper will be held at 90° to the stone's face (*see* Fig 15.2). The edge of the scraper should be stoned until it is smooth and there should be no sign of file marks.

Fig 15.1 The cutting edges of a cabinet scraper.

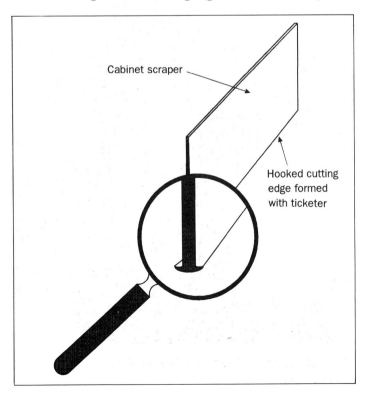

Cabinet scraper

Hooked cutting edge formed with ticketer

Fig 15.2 Stoning the edge of a cabinet scraper. The stone's box lid is used to keep the edge square.

A burr will have been thrown up along both sides of the scraper's edge, which is removed by rubbing it flat on the oilstone. The sides need to be smooth and shiny; the sides of new blades in particular sometimes need work. After this treatment the edge should have corners that are square and sharp. The next part of the process is best done with a special tool known as a **ticketer** (*see* Fig 15.3). The proprietary article, as sold in the tool shop, has a round blade that has been case-hardened. I find that a better tool can be made from an old three-cornered file. First, grind all the teeth off, taking care not to overheat and spoil the steel. The ground file is then polished on progressively finer oilstones. The surface needs to be very smooth and polished, and the corners can be gently rounded. The finished blade needs a good hardwood handle firmly attached. It is possible to make do without a ticketer. The back of a small gouge or the smooth end of a nail punch can also be used.

Fig 15.3 Using the ticketer on the side of the cabinet scraper.

We now come to the part that seems so easy but is so difficult to master properly. I don't know why this is, but after years of working with apprentices, I know that if one perseveres the technique can be mastered. Place the blade that has been filed and stoned flat on the bench. The edge of the scraper needs to be near and parallel to the bench edge. The ticketer is drawn along the scraper with its blade almost flat on the flat side of the scraper. Incline the ticketer very slightly so that it bears on the edge of the scraper. Several passes are made like this on either side, and some pressure is applied to the ticketer, but don't bear down on it too heavily.

Next, the scraper is held on edge with the left hand. Protect the hand by using a piece of rag, or the bottom of the apron. Place one end of the scraper against the bench to keep it firm. With the ticketer in the right hand, draw it firmly up the scraper, bearing on the corner. As the ticketer is drawn along the corner towards you, pull it away from the corner. The purpose is to draw the corner out. This is repeated about four or five times, and each time the angle of the ticketer is increased. The steel the scraper is made from is not hardened, but forming the edge work-hardens the metal. It is not necessary to file the scraper edge every time It needs sharpening. The hook can be reshaped using the ticketer (*see* page 125). When the hook will not form, the edge has to be filed.

A German tool, designed so that a correctly shaped hook is formed every time, does the same job as the ticketer. The tool (*see* Fig 15.4) consists of a block of beech with a metal-lined rebate along one corner. A small part of the periphery of a hardened metal disc projects into the corner of the rebate. The tool is drawn along the scraper. The edge of the disc acts in a similar way to the ticketer. It is normal to sharpen all four edges of the cabinet scraper.

Fig 15.4 A German tool for putting the burr on cabinet scrapers.

Scrapers that are used in a plane have a hooked edge similar to the cabinet scraper. The edge is not filed square but at 45°. A hook is formed by drawing it over with the ticketer. Thickness of shaving is controlled by the amount of scraper that projects from the sole, and not by the size of the hook. The scrapers used in planes are made from thicker material than the normal cabinet scraper.

A very useful square-edged scraper can be made from an old machine hacksaw blade. These are, after all, only strips of high-speed steel. Some large hacksaw machines take blades that are 1½in (38mm) wide. If the teeth are ground off and the edge squared up on the grinder, a very sharp corner is formed. While this tool will not remove the gossamer-like shavings that a properly prepared cabinet scraper does, it will take off polish and varnish while retaining its edge. Many engineering works have hacksaw machines. When the blade is worn out, they are thrown in the skip. If you explain why you require them, you may be given one or two.

KNIVES

The knives used in woodworking can be divided into two broad groups. There are knives used for setting out, and others for shaping wood, such as carving and whittling. Many woodworkers make their own knives, using oddments of steel salvaged from the scrap heap. Old cutthroat razors are a favourite source of fine steel for this purpose. Setting out knives, sometimes known as 'striking knives', come in several patterns. The main difference between them is the way the cutting edge is fashioned. Some have a bevel on both sides of the blade, while others are sharpened with a single bevel, similar to a chisel. I find the single bevel better, as the flat side of the blade fits snugly against a straight edge. For most marking out it is the point of the blade that mainly does the work, so it is here that our attention is focused when sharpening. The shape

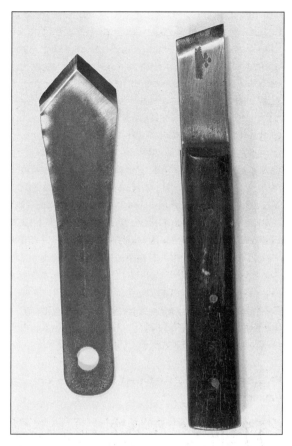

Fig 15.5 Two entirely different setting out knives. The one with the wooden handle is sharpened from both sides, while the other has a bevel on one side only.

of the blade will determine where the point is formed. There are two distinctly different shapes (*see* Fig 15.5). The point on the knife where the end of the blade is splayed at a single angle is usually more acute and difficult to maintain. For most setting out, a single sharpening bevel of 25° is ideal. Grinding is only required when the blade has been damaged, or when modifying the shape of the blade.

The shape of a knife blade used for carving or whittling is usually altered to suit the user. Knife shape is a very personal thing. Most shop-bought knives are reshaped by the owner. This shaping is carefully carried out by grinding. A knife edge can be sharpened with a flat bevel or a gentle round. Chip carving knives usually have a flat bevel. Knives used for whittling and shaping wood perform best with a rounded-off bevel. The round side bears against the wood being shaped and will control the cut. These knives have to be kept very sharp if they are to work properly. As knives tend to wear the stone hollow, a small fine stone is best kept especially for this purpose. It does not matter then if the stone is hollow. A razor strop will be found useful for keeping a keen edge on the knife. The type of four-sided strop that has a stone on one face is the ideal tool (*see* Chapter Six).

The angle of the cutting edge is dependent on the quality of the steel and how the knife is to be used. For heavy cutting the edge has a larger angle than that used for fine cutting. The knife is a tool neglected by many woodworkers, and apart from setting out knives, many workshops do without them all together. To some professionals the use of a knife for shaping wood is frowned upon. Personally, I believe they have a place in the workshop. Knives, once in condition, need some means of storage when not in use. An edge is easily damaged, and some protection is necessary. I find the best storage is a block of balsa. Knives can be stuck into the end grain.

The disposable-bladed surgical scalpel made

Fig 15.6 Ernie Ives' marquetry knife. The parts of the knife are displayed here, with the assembled tool ready for use below.

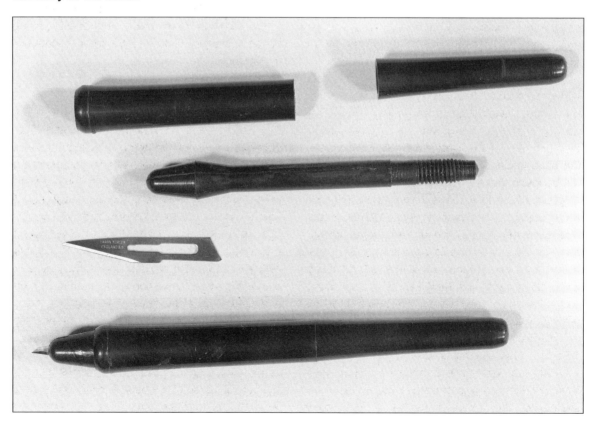

by Swann Morton has found its way into most workshops. Although the blades are disposable, most craftsmen sharpen them. For marquetry these tools have no equal. The blade holder may not be perfect, but the one invented by Ernie Ives and sold by most marquetry material suppliers is ideal (*see* Fig 15.6). The point of the blade gets most use and loses its edge soonest. A small, fine slip stone and a piece of dressed leather can be used to keep a really sharp tip on the blade. The edge on the new unused tool can often be improved by sharpening.

THE AXE

A small hand axe may seem a clumsy tool for use in the workshop to the uninitiated. However, in the hands of a craftsman it is an important item. Like all other woodworking tools it will only perform properly when sharp. There are many patterns of axe, but there are only two features that affect the sharpening technique. A side axe is sharpened with one face flat and the other bevelled, similar to a chisel, whereas the normal axe has a bevel on both sides. I must emphasize that we are talking here about hand axes, not the variety that have long handles and require two hands to control them. Even so, there are a number of different sizes and weights. The size and weight of the head influences the method of sharpening. It will be found that the smaller and therefore lighter heads are best held in the hands and rubbed on the oilstone. The big and heavy heads are better placed on the bench edge and the stone rubbed on them. The side axe is sharpened with a flat bevel. The angle of this bevel to the flat side is lower than most woodworking tools, at around 22°. Most woodworkers hone this bevel along the cutting edge at a slightly steeper bevel of 25°. Rub the flat side on the stone to remove any burr. Take care to keep it flat on the stone.

The axe head that is sharpened from both sides does not have pronounced flat bevels when sharpened. The cheeks of the head are rounded

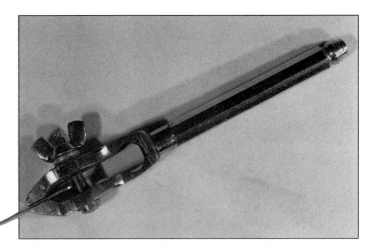

Fig 15.7 Jewellers' hand vice being used to hold small blades for sharpening.

gently into the sharpening angle. The edge is ground from both sides at an included angle of 25° or slightly less. The cutting edge is honed from each side of the head, being careful to remove any burr. Only the very edge of the blade is honed. An angle of 30° will be found suitable for most uses.

THE ADZE

There are several different designs of adze. Each different branch of the trade seems to have its own favourite design. Apart from the shape of the head there are two distinct designs: the adze that is held with two hands, and the hand adze that is held in one hand. Despite these differences the sharpening is identical for all heads. The head is placed on the bench with the handle pointing vertically up to the ceiling. The cutting edge is near and parallel to the bench edge. An oilstone is rubbed on the inside bevel. The bevel is usually kept flat at one angle, i.e. there is no separate grinding and honing angle. When the inside bevel has been honed, the head is turned over. The outside of the head is now worked with the stone. The curved shape of the head is continued right to the cutting edge; there is no bevel.

**Fig 15.8
Cutting
gauge blade.**

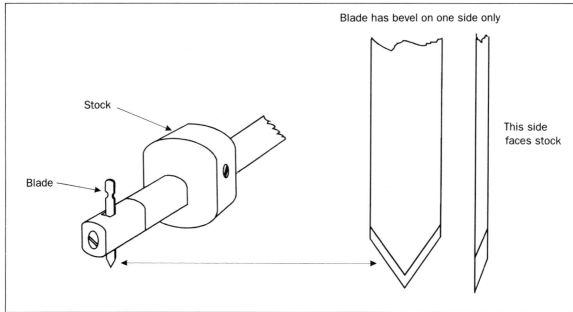

Blade has bevel on one side only

Stock

Blade

This side
faces stock

CUTTING GAUGES

This is a tool whose sharpening is often overlooked, yet the blade from a cutting gauge needs frequent attention. Take the blade from the gauge for sharpening. It will be found small and difficult to hold. I have a small jeweller's hand vice that is ideal for holding these small blades (*see* Fig 15.7). The blade is sharpened with bevels on one side only; the other side remains flat. The cutting end of the blade has a double splay (*see* Fig 15.8). Some cheap tools have soft blades which will not hold a sharp edge. A new cutter can easily be made from an old broken hacksaw blade.

A large-headed adze is put on the handle so that the head tightens as the adze is used. This can be likened to the joiner's mallet. By tapping the end of the handle the head can be removed. This makes the tool easier to sharpen.

SCISSORS

While a pair of scissors can hardly be called a woodworking tool, there is at least one pair in most workshops, and these will eventually need sharpening. Scissors cut by a shearing action. The edges of the blades are sharpened at an angle of 60°. Some people consider that an edge straight from the grinder is good enough. However, the tool benefits from having the bevel rubbed on an oilstone after grinding. Some scissors have the blades held together with a nut and bolt, and it is easier if they are taken apart before sharpening. The end of the bolt is often peened over to prevent it becoming slack in use. The peen can be removed with a file, and peened again with a small hammer after assembling.

SPOKESHAVES

The blade from a metal spokeshave can be likened to a small plane blade, and is sharpened in exactly the same way (*see* Chapter Eight). The problem with the spokeshave, however, is that the blade, being small, is difficult to hold in the hands while honing. Most craftsmen make a tool to hold these blades (*see* Fig 15.9).

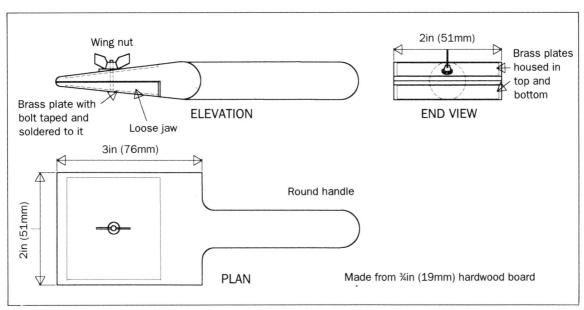

Fig 15.9 Holder for metal spokeshave blades.

Wing nut

Brass plate with bolt taped and soldered to it

Loose jaw

ELEVATION

3in (76mm)

2in (51mm)

Round handle

PLAN

2in (51mm)

Brass plates housed in top and bottom

END VIEW

Made from ¾in (19mm) hardwood board

The wooden spokeshave has a solid forged cutter, which has much to recommend it. The cutter is almost chatter-proof when compared to the metal spokeshave. Because of the two prongs on the cutter securing it to the spokeshave, it cannot be sharpened on the flat face of a bench stone. Stand the stone on edge. Hone the inside bevel, which is the part between the two prongs, taking care to keep the original bevel. It is very easy to give the centre of the blade more attention than the ends; avoid this. Make sure the cutting edge is kept straight. The face of the cutter is rubbed flat on the stone. It may be found that the face of the cutter is gently rounded from one end to the other. This shape should be carefully maintained.

DRAW KNIVES

Here we have another tool that is held while the stone is moved to hone it. The edge has a ground bevel of 25°. The method of grinding the tool will depend on the machine being used (*see* Chapter Three). The blade has a bevel on one side only, the other being flat. The tool to be honed is placed in a block of wood that has an angled groove worked in one edge (*see* Fig 15.10). The stone is now passed along the edge from end to end. A honing angle of 30° is suitable for most work. The block used to hold the tool can have an angle worked on the back edge on which the stone will rest to help maintain the honing bevel. Take care to keep the hands clear of the edge being honed. It is very easy to cut yourself.

Fig 15.10 Draw knife in block for sharpening.

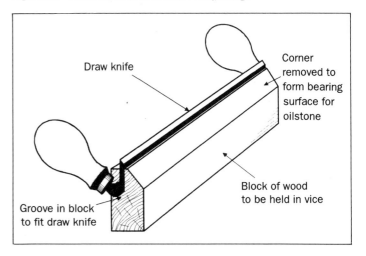

Draw knife

Corner removed to form bearing surface for oilstone

Groove in block to fit draw knife

Block of wood to be held in vice

CHAPTER 16

SHARPENING JAPANESE EDGE TOOLS

WATER STONES

Over the past decade Japanese tools have found their way into many Western workshops. Some Western craftsmen use them almost to the exclusion of their traditional tools. Others have tried and rejected them. However, I know of only one person who, after acquiring a Japanese water stone, discarded it. With this one exception, everybody I know who has used these stones speaks highly of them. Before explaining how to use water stones, I must point out that Japanese tools vary from traditional Western tools in one major way. In the West we buy a tool, sharpen it and it is ready to use. Apart from saws, all Japanese tools need some preparation before they are in working condition. A Japanese *shokunin* (craftsman) would consider it an insult if somebody else prepared his tools. This is not known by many Western purchasers of Japanese tools. Therefore, the tool is not prepared properly, performs poorly, and is rejected. What a shame this is. The very best Japanese tools are still handmade, often by a craftsman working on his own. The skills he uses are those passed on by the Samurai sword makers. When the Samurai were outlawed at the beginning of the Meiji era in 1868, some *katana-kaji* (swordsmiths) turned to making woodworking tools. This is why top quality Japanese tools are so superior to anything made in the West. Each blacksmith making tools by hand has his own secret way of treating the steel. Because of this, the tools made by a *shokunin*

have particular characteristics that distinguish them from those made by other craftsmen.

A STONE FOR EACH BLADE

The Japanese woodworker goes to great lengths to match his sharpening stone to the tool. This is not an easy task in Japan where there are many fine quality natural water stones to be had. Stones vary from quarry to quarry, and each stratum in the same quarry produces a stone that has an individual characteristic. Toshio Ôdate in his book *Japanese Woodworking Tools: Their Tradition, Spirit and Use*, says that the matching of a blade with a stone is like a marriage of the perfect bride and groom. Not surprisingly, the Japanese craftsman may have as many as 30 stones in his tool kit. We in the UK do not have access to many of these natural stones. Those that do find their way here are very expensive. There are, however, man-made Japanese water stones. Even if the natural stone is available, you will probably be better off buying the man-made stone since it is of known quality and consistency and is also a fraction of the price of a natural stone. Man-made stones are graded by their grit size.

The Japanese craftsman's attitude to his tools is very different to that of the average Western woodworker. Unless the stone and the technique used to sharpen a tool produced the best edge the tool was capable of, it would be an

Fig 16.1 Japanese water stones in a trough. The stones not in use are inverted and are thus kept under water.

insult to the tool and the man who made it. The relationship of the *shokunin* to his tools is almost spiritual. Perhaps this is one of the reasons why top Japanese craftsmen are held in such high esteem in their country.

Both natural and man-made Japanese water stones are renowned for their cutting speed. This is the thing that one notices when using the stone for the first time. There are two types of abrasive used to make the stone: synthetic material or crushed natural stone. However, stone composition and manufacturing techniques are kept a closely guarded secret by most of the manufacturers. It is 15 years since I bought my

first Japanese water stone, and I would now be lost without it. These stones work wonders on Western tools. My tools were never as sharp before that happy event as they are now. I would go so far as to say, provided I could keep my water stones, all the others would not be greatly missed.

Grades of Stone

The two main brands of water stone available in this country are King and Sun Tiger. They are obtainable in the following grades: 200, 700, 800, 1000, 1200, 4000, 6000 and 8000. Size is very similar to that of a Western bench stone: 8in long and 2in to 2½in wide (203mm by 50–63mm). The two finest grades in the King range are only ½in (13mm) thick. To give these stones added strength they are mounted on a wooden base. After being in use for a few months the adhesive gives up and the stone and wood part company. The stone then needs to be mounted properly in a box (see below).

My sets of stones are in two troughs; the main set comprise 700, 800, 1200, 4000, 6000 and 8000 grades, and these are used for all tools. The second set (1200, 4000, 6000, and 8000 grades) are used for wide plane irons only. When a stone in the first set becomes hollow, it is flattened and exchanged with a stone from the second set. In this way my plane blade sharpening stones are always flat.

You may be interested to know the Japanese names for the different grades of sharpening stone. These are: *ara-to* (coarse stone, 700 to 800); *naka-to* (medium stone, 1000 to 1200); *awase-do* (finishing stone, 4000 to 8000). The 8000 is often called a polishing stone. The speed that these waterstones cut at compared to oilstones is phenomenal.

As well as these stones there is another tool in the Japanese sharpening armoury: the *uraoshi* (steel flattening plate) (*see* pages 136–7) which is a plate of soft steel that is used with silicon carbide grit to flatten the face of tools. I have found that a diamond sharpening plate is more effective when a tool needs much work on it. However, one needs the *uraoshi* to finish the flat on the face of chisels and plane irons.

Dressing the Stone

Japanese water stones are soft and are therefore easy to dress. This can be done in several ways. I use a sheet of aluminium oxide paper taped flat on a piece of glass. The stone is allowed to dry out overnight before it is rubbed flat. It is possible to dress the stone while it is still loaded with water, but the slurry clogs up the abrasive paper. Some craftsmen keep their stones flat by rubbing

the faces of the stones together. This is done with three stones. They are each rubbed alternately against the other two stones. When each shows contact over its entire surface, they are all flat. Because the stones are soft, care must be taken when sharpening on their surface. It is very easy to dig the edge of the tool being sharpened into the surface of the stone. This is one reason I use a honing guide.

Stone Maintenance

Like all our tools, the water stone must be looked after and cared for properly. Because of the way the stone works, it relies on a copious supply of water, not only on the stone's surface but also throughout its body. A dry stone immersed in water will absorb an astonishing amount. It must never be used dry. Being soft, the stone is continuously wearing away, and particles freed from the surface of the stone form a **slurry** on its surface. This slurry acts as a sharpening medium. There is a small stone called a *nagura-to* which is rubbed on the surface of the stone before sharpening. This abrades the surface and forms a slurry. Stones are kept immersed in water when not in use, ensuring that they remain waterlogged (*see* Fig 16.1). If you only have one or two water stones they can be kept in the plastic containers that bulk ice cream comes in. You can prevent it becoming smelly by using a very small amount of household bleach.

A proper trough in which to keep the stones can be made. Teak (*tectona grandis*) is the best wood to use, because of the ever-present water. Fig 16.2 shows a section through the trough with a stone mounted in a wooden base. The wooden base is so shaped that when the stone is not in use it is turned upside down; it is then below the surface of the water. Slurry from the surface of the stone falls off and gravitates to the bottom of the trough. This abrasive slurry can be dug out from time to time. It is worth saving as a useful fine abrasive.

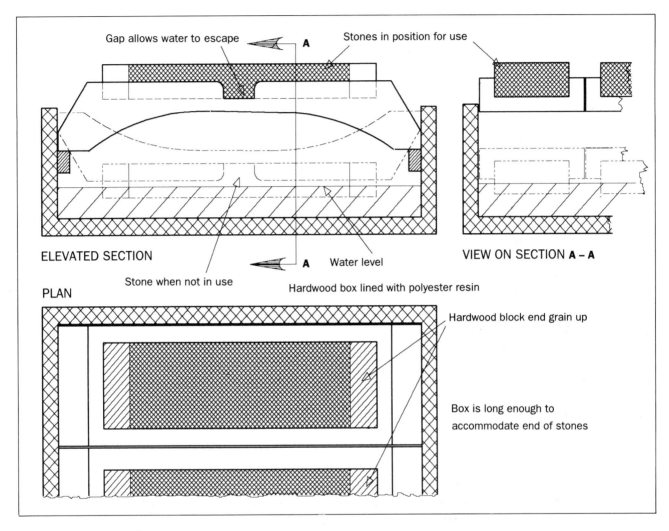

Fig 16.2 Water stone box.

JAPANESE CUTTING TOOLS

The blades of most Japanese cutting tools are made by laminating a thin piece of high-carbon steel on to soft steel. There was a time when Western tools were made in a similar way, before the days of mass production. The best hand-forged Japanese tools are made from soft steel that was smelted before 1900. This has some impurities in it, such as silica particles, which are refined out of modern steels. The Japanese

blacksmith obtains this steel from scrapped items such as old ship's boilers, anchor chain and suchlike. The correct scrap steel is hard to find, and the process of working it is not commonly known. To make a blade, a suitable bar of steel is forged by using heat and beating it to shape on an anvil. This heating and beating has a beneficial effect on the structure of the steel. When the blacksmith is satisfied with the shape, he places the billet back in the forge and heats it to white-hot temperature. The billet is removed from the forge, tapped to remove any slag, then flux is added to the surface to be welded. The thin piece of high-carbon steel is placed on the white-hot bar by the blacksmith using his bare fingers. He hammers the steel, bending it down the sides of the billet. The assembled metals are placed back in the forge and heated to welding temperature. The pieces of metal are then hammered together

Fig 16.3 The back of a Japanese plane iron. The hollow can clearly be seen; only the very edge is flattened.

at a temperature around 950°C.

The block formed in this style is forged to the shape required. The way these processes are done directly affects the quality of the final tool. The technique is so difficult that it takes an apprentice about ten years to master the necessary skill. After this there is still heat treatment and grinding to be done. The way the individual blacksmith carries out these processes varies, and they are mostly carefully guarded secrets. One smith packs the tool in river mud before putting it in the forge to heat treat it. Certain minerals from the mud percolate into the surface of the steel, imparting particular characteristics. I have had the hardness of several Japanese chisels tested. They are hardened to 65° Rockwell C. European chisels from our major manufacturers are between 58° and 60° Rockwell C; any harder than this and the edges chip.

You will appreciate that a tool created with this level of skill and knowledge deserves your respect when you come to use it. The sharpening technique differs from Western methods. First, the blade does not have a flat face as Western tools do. There is a hollow in it, hammered in by the blacksmith when forging the tool. The purpose of the hollow is to allow the face adjacent to the cutting edge to be maintained perfectly flat. Japanese call this flat area the *ura,* and it is best appreciated by inspecting a Japanese plane blade (*see* Fig 16.3). Some chisels have several hollows, yet the sharpening technique is the same for all blades. A person who does not understand the tool would, on looking at the face, assume the life of the tool to be very short. Surely it is unuseable once the cutting edge has been sharpened a number of times and the flat portion of the blade's face meets the hollow? Not so, thanks to the *uraoshi* (*see* page 134). The flat of the blade is rubbed on the *uraoshi* using Carborundum powder, starting with a coarse grade of about 50 grit and gradually using finer grades until a new flat is formed. Finally, the fine slurry from the polishing stone is used on the *uraoshi.*

Fig 16.4 The blade of a Japanese chisel. The hard steel around the cutting edge can clearly be seen.

The face of the blade must be polished until it appears as if it has been chromium plated. When this process is finished, it will be found that the sides of the chisel are sharp enough to cut the fingers. Take a small stone and just remove these edges.

The bevel is the next part to receive attention. The Japanese cutting tool does not have two bevels as is customary on our Western blades; there is only the honing angle. The main bulk of the blade is made from soft steel that is easily abraded away. This gives a very wide flat surface that can easily be held flat on the stone. The bevel is rubbed on progressively finer grades of stone. It will be found that the 8000 will impart a brilliant shine to the surface. The high-carbon steel face can clearly be seen when the bevel is inspected. There is quite a difference in colour between the two grades of steel from which the tool is made (*see* Fig 16.4). The face of the blade is rubbed flat on the finishing stone. There is no reason to rub the face on any of the other stones,

as this would only incur more work polishing the abrasion marks out.

PLANE IRONS, SPECIAL TREATMENT

The flat part of the plane iron near the cutting edge becomes very narrow from constant sharpening, and the time will come when some very special treatment is called for. The face of the iron is placed on an anvil (or else a block of hardwood), and the bevel is gently tapped with a hammer (*see* Fig 16.5). The objective is to bend the metal forward to form a new flat. Once sufficient metal has been brought forward, the iron is flattened on the *uraoshi* as described above. There are special tapping tools sold in Japan for this purpose, but unfortunately they are not available here. The hammer best suited to the purpose is the square-headed Japanese type. The process should be treated delicately; working along the whole length of the bevel, try to bend the metal forward in gentle stages. Make sure that all the hammering is done on the soft steel, and under no circumstances strike the hard steel laminated on the face, as impatient blows from the hammer can easily crack it. The different metals can clearly be seen on a newly honed bevel.

When sharpening the iron on the coarse and medium stones, they are rubbed until a burr appears on the edge. This burr is usually so small that it cannot be seen, but it is possible to detect it by drawing the ball of the thumb down the face of the tool to the cutting edge. The finishing stones do not produce a burr. As the abrasive particles of the finishing stones are so fine, there is no advantage to be gained by using a strop after honing on them. When sharpening Japanese tools there is no need for a grinding machine.

Japanese plane irons need to be held slightly differently than Western irons when they are being honed. The short Japanese iron needs holding so that pressure is placed evenly across the cutting edge. The thumbs of both hands are placed on the underside of the iron, and the fingertips of the left hand are placed across the iron pointing towards the cutting edge. Only the forefinger of the right hand is on the face of the iron; the remaining fingers are curled up against the palm.

Fig 16.5 The back of a Japanese plane iron being flattened by tapping with a hammer.

CHAPTER 17

SHARPENING AND SETTING JAPANESE SAWS

THE DIFFERENCE

Japanese saws can be divided into two classes: *kataba* (single-edged) and *ryoba* (double-edged). *Kataba* saws are made to perform many tasks. *Ryoba* saws have rip teeth on one edge and crosscut on the other.

Japanese saws are completely different from Western saws. While a Western saw cuts on the push stroke the Japanese saw cuts on the pull stroke. This idea is fundamental to the characteristic of the different tools. A saw that is pushed must be thick and rigid, or it will bend rather than enter the cut. A saw that is pulled is in tension and remains straight, and can therefore be made from much thinner metal. This has one tremendous advantage: a saw must cut a kerf that allows the blade to work freely, and the width of the kerf is made by removing wood. This requires energy, and for the same length and depth of cut, a Japanese saw – being made from thinner metal – requires less effort than a Western saw.

The steel it is made from is much harder than that of the Western saw, and its teeth are easily broken. Japanese saws are not specified by teeth per inch: the size of their teeth is dependent on the length of the saw. I find the thin saw more accurate in use than the thick one. The difference can be likened to drawing with a very sharp pencil, compared to using a crayon.

THE TEETH

A Japanese ripsaw has teeth that are in some ways similar to those on a Western ripsaw. The tips of the teeth are filed square so that they are chisel-like. The rake and other angles of the tooth are different however. The size of teeth are graded from small at the heel to large at the toe. Some small saws do not have these graded teeth, all their teeth being the same size. There is yet another difference between Japanese and Western saws: the Japanese use different ripsaws for softwood and hardwood (*see* Fig 17.1); the softwood teeth are more pointed than those used for hardwood. If the softwood teeth were used on hardwood they would dig in and the saw would be inclined to chatter rather than cut. The larger angle on the hardwood saw also adds strength to the teeth.

The Japanese crosscut has a totally different tooth configuration to Western saws, with each tooth resembling a separate knife. The teeth do not vary in size as the ripsaw's do; all the teeth from heel to tip are the same. Here again there is a difference between the saw used for cutting softwood and that for sawing hardwood. Some Japanese cross cut saws do not have set on the teeth (*see* Fig 17.1)

As well as the saws used for ripping and crosscutting, there are also saws for cutting at an angle to the grain. These saws have *ibara-me* teeth for softwood and *nezumi-ba* teeth for

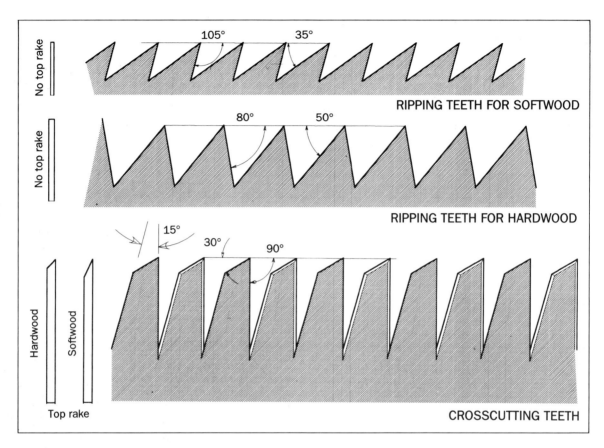

Fig 17.1 Tooth configuration on Japanese saws used for ripping and crosscutting.

hardwood (*see* Fig 17.2). In addition to this there is yet another tooth configuration. This is a recent development, probably to reduce the number of saws a woodworker requires. It is used on general purpose saws as it is effective both across and down the grain. There is a series of crosscut teeth – usually between six and ten – followed by two ripping teeth. The ripping teeth are fractionally shorter than the crosscut teeth, and they clear the bottom of the cut. This tooth configuration is known as *ikeda-me* (*see* Fig 17.3).

SETTING

Compared to a Western saw, the blade of the Japanese saw is very brittle, and it is easy to snap

a tooth out of the blade. In Japan, the sharpening and setting of a saw are normally entrusted to a craftsman who specializes in the task. The Japanese doctor, known as a *metate-shokunin*, not only sharpens and sets saws but he also deals with buckled saws. Where an expensive tool has been damaged, he will cut a completely new set of teeth. Unfortunately there is nobody in this country with this sort of skill. Owners of Japanese saws have to service them, and acquiring this skill can be expensive in spoilt saws. Consequently it is best to develop the skill on a low-cost tool.

The Japanese use a tool called a *mefuri* to apply the set to the saw tooth; in its simplest form this is a piece of metal with a slot in it that fits over the tooth to be bent. This tool works the same way as the old-fashioned manual saw set used in the West. A pliers saw set, made to set Western saws, works well on the Japanese saw. It is important that a good quality pliers saw set is used. Some teeth that are to be set are very small, and the plunger in the saw set must be narrow or it will not fit on just one single tooth. Because of the very thin gauge of the saw and the brittleness of the steel, it is best to apply the smallest amount of set that you think is required. If this is insufficient, more set can be given. When the correct amount of set has been established, the number the anvil is set at can be recorded, and the next time the saw needs setting it will be possible to set the pliers up correctly at the start.

SHARPENING

Each separate part of the saw tooth on a Japanese saw is individually named (*see* Fig 17.4). The leading edge of the tooth is the *shita-ba*, and the trailing edge the *uwa-ba*. The top of the crosscut tooth is the *uwa-me*. The angle of the cutting edge is the *kiri-ba*.

The files used to sharpen these saws are totally different from any Western file (*see* Fig 17.5). There are two principal files: the *surikomi-yasuri* (cutting down file), used to maintain the

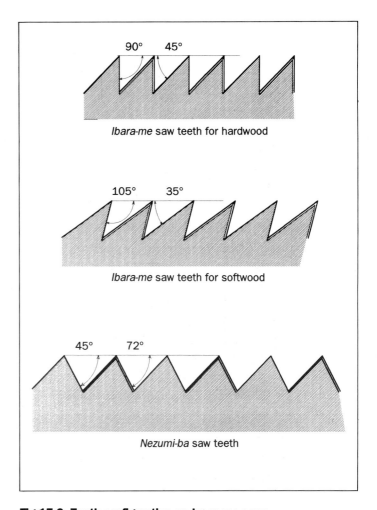

Ibara-me saw teeth for hardwood

Ibara-me saw teeth for softwood

Nezumi-ba saw teeth

Fig 17.2 Tooth configuration on Japanese saws used for cutting at an angle to the grain.

Fig 17.3 *Ikeda-me* saw teeth.

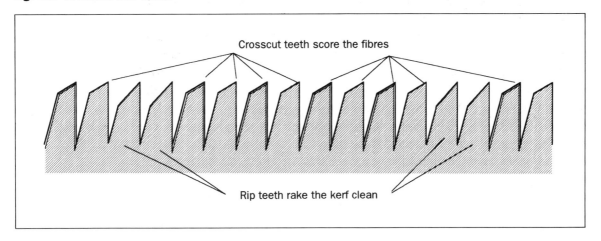

Crosscut teeth score the fibres

Rip teeth rake the kerf clean

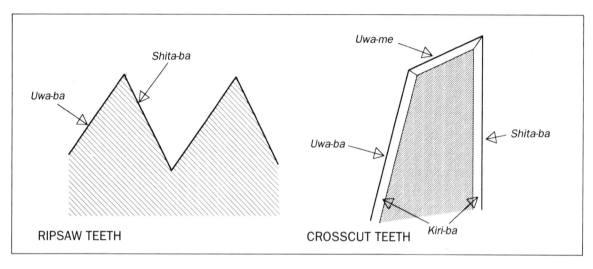

RIPSAW TEETH

CROSSCUT TEETH

Fig 17.4 The Japanese names for the different parts of the saw tooth.

Fig 17.5 Japanese saw files.

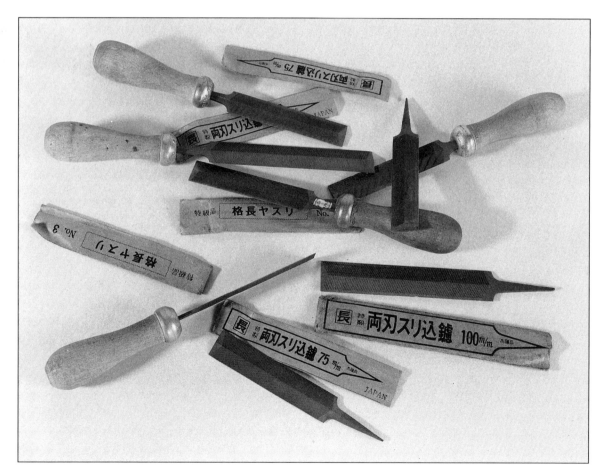

shape and depth of the tooth, and the *hatsuke-yasuri* (cutting edge file), used to sharpen the cutting edge of the teeth (this file cuts on both the push and the pull stroke). A third file called an *uwa-me-yasuri* is needed, made from a *hatsuke-yasuri*. The teeth on the edge of the *hatsuke-yasuri* are polished off and slightly rounded on a fine oilstone. I find that a file that has been used to sharpen a saw once or twice loses the cut along the edge and, rather than spoil a new file, it is one of these files that I use to make into *uwa-me-yasuri*. The Japanese use a slender piece of pine for a file handle. This is straight from the tree with the bark still on it. The wood is very soft and

supple, so the tang of the file can be pushed into the end grain quite easily. I have tried this but have not liked the feel of the bark and the sticky sap. Perhaps I have the wrong pine tree. I turn up a suitable handle in the lathe, which can be transferred to each file as it is needed. A plastic handle supplied with a double-ended Western saw file can be used if you have a spare one.

Like the Western saw, the Japanese saw needs holding firmly while it is filed. I use a saw sharpening block for this, which works fine. Should you wish, you can make the appliance the Japanese use. Fig 17.6 shows one in detail.

Secure the crosscut saw in whatever holding

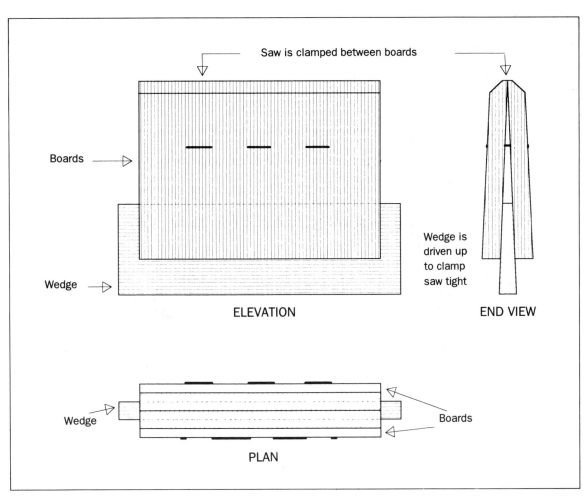

Saw is clamped between boards

Boards

Wedge

ELEVATION

Wedge is driven up to clamp saw tight

END VIEW

Wedge

Boards

PLAN

Fig 17.6 Japanese-style saw sharpening block.

device you are using. Only the teeth and a very small amount of the blade should project. Take the *hatsuke-yasuri* and place the feather edge down in the gullet. Push the file flat against the front face of the leading tooth. The file should adopt the original angle at which the tooth was

Fig 17.7 When sharpening saws with small teeth, some form of magnification will be found helpful. Those seen here can be worn in addition to one's normal spectacles.

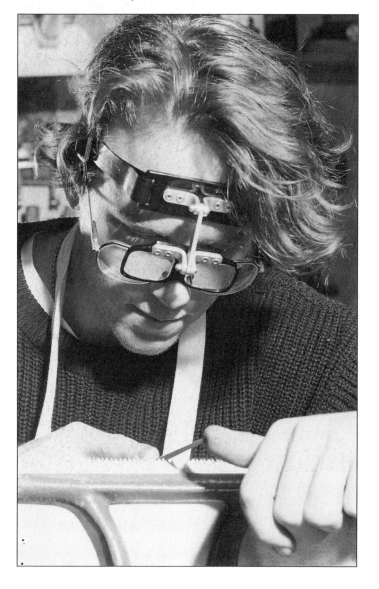

filed. Using a light hand and very little pressure, pull the file once along the bevel of this tooth. The saw blade is thin and the file has fine teeth, so a delicate touch is called for. Do not use the file as you would a Western tapered saw file. Miss a tooth and repeat the process; skipping each alternate tooth, work down the whole length of the blade, then turn the saw round and, from the other side, repeat the process on the teeth missed the first time. The same operation is done on the back of each tooth, meaning that you must go along the saw four times. Next, the top of the teeth need filing, and for this the *uwa-me-yasuri* is used. The edge that was rounded on the oilstone is allowed to rub against the tooth next to the one being filed, which is why the corner was rounded. Keep the file angle correct, and carefully pull it towards you. Note that the file is pulled, not pushed as we would a Western one. The tips of the teeth usually only need half a stroke as they are very small. Each alternate tooth is treated in this way, and the saw is then turned round and the other teeth filed. You should have gone six times down the full length of the saw by now, filing each alternate tooth. The sharpening process takes time, and requires good light. I find that with the fine teeth on my *dozuki-nokogiri* (similar to a dovetail saw) I need some form of magnification. I use a pair of magnifying glasses that are fixed to a headband (*see* Fig 17.7).

The Japanese ripsaw is sharpened in a similar way to the crosscut. It is much easier to sharpen, having only two bevels to each tooth. However, the front and the back of each tooth are filed separately. The teeth do not have the 60° angle which is so convenient on our Western saws.

CARE OF THE SAW

The Japanese saw will have a saw cover on it when it is bought. On a cheap saw this will probably be made of plastic. The better saws have a heavy denim or canvas cover. This cover should be kept on the saw when it is not in use. However,

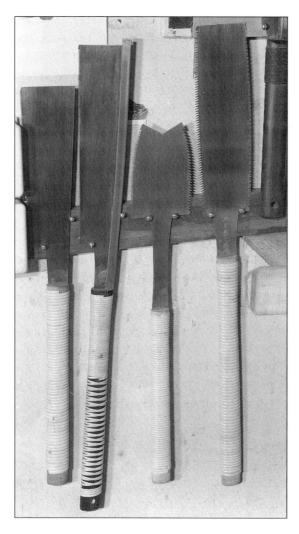

Fig 17.8 Japanese saws are best stored in the workshop by hanging them on the wall.

Fig 17.9 Japanese saw with its cover.

in the workshop it can be inconvenient to put on and remove the cover every time, and the saws can be hung on screws in a board (*see* Figs 17.8 and 17.9). The Japanese saw is tempered so that the steel is hard, and it stays sharp longer than a Western saw. However, the sharpening process, when it has to be performed, takes much longer. When a Western saw runs into a nail or other hard obstacle in a piece of wood, the result can be put right in half an hour. The Japanese saw in the same accident may lose half a dozen teeth. It should therefore never be exposed to the risk, if it can in any way be helped.

SUPPLIERS

Most items required for sharpening woodworking tools are obtainable from retail tool shops. Some of the more difficult items to find are obtainable from the following places:

Diamond Sharpening Systems

Starkie & Starkie Ltd,
118 South Knighton Road, Leicester LE2 3LQ
Tel: 0533 703212

Diamond Machining Technology Inc.,
85 Hayes Memorial Drive, Marlboro, MA
01752-1892, USA
Tel: 508 481 5944

Eze-lap Diamond Products,
Box 2229, Westminster, CA 92683, USA
Tel: 714 847 1555

Tilgear,
Bridge House, 69 Station Road, Cuffley, Herts
EN6 4TG
Tel: 0707 873545

Grinding Machines

Black & Decker,
Westpoint, The Grove, Slough, Berkshire SL1 1QQ
Tel: 0753 511234

Tormek AB,
Box 152, Lindesberg, Sweden

Record Power Ltd,
Parkway Works, Sheffield
S9 3BL
Tel: 0742 434370

Grinding Wheels

Williams Wheel (Abrasives) Ltd,
Unit 1C, Mount Industrial Estate, Stone, Staffs
ST15 8LL
Tel: 0785 813828
(This company also makes oil and slip stones.)

Robert Wilcox (Abrasives) Ltd,
19 Desborough Avenue, High Wycombe, Bucks
HP11 2RS
Tel: 0494 530533

P.B.R. Abrasives (W'ton) Ltd,
Wolverhampton Street, Willenhall, West Midlands,
WV13 2NF
Tel: 0902 368624

Japanese Water Stones

Thanet Tool Supplies,
Romney House, Elwick Road, Ashford, Kent
TN23 1PG
Tel: 0233 636304

Safety Equipment

Racal Health & Safety Ltd,
Beresford Avenue, Wembley, Middlesex HA0 1QJ
Tel: 081 902 8887

BIBLIOGRAPHY

Purpose Made Joinery Peter Brett, Hutchinson

Wood Machining Nigel S. Voisey, Stobart

The Practical Woodworker Bernard E. Jones, Waverley

The Stanley Plane Alvin Sellens, Early American Industries Association

Wide Bandsaws Arthur Simmonds, Stobart

The Technique of Furniture Making Ernest Joyce, Batsford

Planecraft Hampton and Clifford, Hampton

Cabinetmaking and Millwork John L. Feirer, Chas A. Bennett Co., Inc.

Joinery and Carpentry Corkhill and Lowsley, New Era

Woodwork Tools William Fairham, Evans Brothers

Tools for Woodworking Charles H. Hayward, Evans Brothers

Practical Woodcarving and Carving William Wheeler and Charles H. Hayward, Evans Brothers

The Craftsman Woodturner Peter Child, G. Bell & Sons

Machine Woodworking Technology for Hand Woodworkers F.E.Sherlock, Newnes

The Technology of Woodwork and Metalwork Norman R. Rogers, Pitman

Wood Carving, Design and Workmanship George Jack, John Hogg

Tool Grinding & Sharpening Handbook Glenn D. Davidson, Stirling

The Grinding Machine Ian Bradley, MAP

METRIC CONVERSION TABLE

INCHES TO MILLIMETRES AND CENTIMETRES

MM = Millimetres CM = Centimetres

INCHES	MM	CM	INCHES	CM	INCHES	CM
⅛	3	0.3	9	22.9	30	76 2
¼	6	0.6	10	25.4	31	78.7
⅜	10	1.0	11	27.9	32	81.3
½	13	1.3	12	30.5	33	83.8
⅝	16	1.6	13	33.0	34	86.4
¾	19	1.9	14	35.6	35	88.9
⅞	22	2.2	15	38.1	36	91.4
1	25	2.5	16	40.6	37	94.0
1¼	32	3.2	17	43.2	38	96.5
1½	38	3.8	18	45.7	39	99.1
1¾	44	4.4	19	48.3	40	101.6
2	51	5.1	20	50.8	41	104.1
2½	64	6.4	21	53.3	42	106.7
3	76	7.6	22	55.9	43	109.2
3½	89	8.9	23	58.4	44	111.8
4	102	10.2	24	61.0	45	114.3
4½	114	11.4	25	63.5	46	116.8
5	127	12.7	26	66.0	47	119.4
6	152	15.2	27	68.6	48	121.9
7	178	17.8	28	71.1	49	124.5
8	203	20.3	29	73.7	50	127.0

ABOUT THE AUTHOR

Jim Kingshott was born in 1931 in Surrey, where he still lives. He is a professional cabinetmaker who served his apprenticeship in the late 1940s and gained experience in a wide variety of posts, ranging from undertaking to aircraft construction; in more recent years he has been involved in the training of apprentices.

He has been a regular freelance contributor to woodworking magazines for a number of years. He has also made a study of the history of woodworking and the tools associated with it. This led to his first book, *Making and Modifiying Woodworking Tools*, published by GMC Publications in 1992. Since then he has written *The Workshop* (1993), also published by GMC Publications.

INDEX

Numbers in figure references indicate firstly chapter and then figure number